DEFENSIVE ESTIMATING

Protecting Your Profits

DEFENSIVE ESTIMATING

Protecting Your Profits

William Asdal, CGR

BuilderBooks.com
BOOKS THAT BUILD YOUR BUSINESS

Defensive Estimating: Protecting Your Profits

BuilderBooks, a Service of the National Association of Home Builders

Christine B. Charlip	Publisher
Doris M. Tennyson	Senior Editor
Courtenay S. Brown	Book Editor
Torrie Singletary	Production Editor
Granville Woodson	Cover Design
Circle Graphics	Composition
Midland Information Resources	Printing
Gerald M. Howard	NAHB Executive Vice President and CEO
Mark Pursell	NAHB Senior Staff Vice President, Marketing & Sales Group
Lakisha Campbell	NAHB Staff Vice President, Publication & Affinity Programs

Printed in the United States of America

24 23 22 21 3 4 5

Library of Congress Cataloging-in-Publication Data

Asdal, William.
Defensive estimating : protecting your profit / William Asdal.
p. cm.
Includes bibliographical references and index.
ISBN-13: 978-0-86718-620-8
ISBN-10: 0-86718-620-8
1. Building—Estimates. 2. Dwellings—Remodeling. I. Title.

TH435.A85 2006
692'.5—dc22

2006047471

For further information, please contact:

National Association of Home Builders
1201 15th Street, NW
Washington, DC 20005-2800
Visit us online at www.BuilderBooks.com.

eISBN 978-0-86718-703-8

To

Doris Asdal

who educated me to cover my back.

Contents

LIST OF ILLUSTRATIONS viii

ACKNOWLEDGMENTS xi

ABOUT THE AUTHOR xiii

FOREWORD xv

PREFACE xvii

1 Understanding the Critical Number in an Estimate—Profit 1

2 Establish the Company Profit Number Based on Your Income Needs 9

3 Competencies for Building and Remodeling Success—Estimating 19

4 Spreadsheet Estimating 29

5 Seven Ways to Get the Numbers 49

6 Using Retail Pricing at Every Line 61

7 Minimizing the Workload 67

8 Materials Costs and Tracking 75

9 Estimating and Tracking Production 85

10 Financial Analysis: Estimating the Cash Flow 93

11 Defending the Profit Line in Your Building Estimate 99

12 Defending the Profit Line in Your Remodeling Estimate 115

NOTES 133

SELECTED BIBLIOGRAPHY 135

INDEX 136

List of Illustrations

Chapter 1

1.1 Title Page of Business Plan 2
1.2 Work Flow in Estimating 3
1.3 A Profitable Day Building to Net Worth 5
1.4 Strategic Planning Cycle 6
1.5 The Reason to Be in Business 6
1.6 A Remodeling Company Supports a Change in Lifestyle 7

Chapter 2

2.1 Passive Versus Active Income 11
2.2 Passive Versus Active Income—Financial Freedom 12
2.3 Personal Financial Statement 13
2.4 Gross Sales Planned by Income Needs 14
2.5 Sample Budget for a Small-Volume Remodeling Firm 16
2.6 Scalable Markup 18

Chapter 3

3.1 How Estimating Fits into the Competencies for Overall Success 20
3.2 Educational Resources for Skills Development 23

Chapter 4

4.1 Short Form Spreadsheet 32
4.2 Converting the Estimate to a Proposal 38
4.3 Converting the Estimate to a Draw Schedule 39
4.4 Converting the Estimate to a Statement 40
4.5 Full Spreadsheet 41

Chapter 5

5.1 Entering Formulas 53
5.2 Day Rate Cost Estimates Versus Assembly Cost Estimates 55
5.3 Quantity Takeoff for Windows 59
5.4 Takeoff Entries 60

Chapter 6

6.1 The Path to a Street Price for Roofing 62
6.2 Pricing Strategy: Beating the Estimate at the Line Item Level 63

Chapter 7

7.1 Some Sales Criteria Limit the Jobs You Take 68
7.2 Prequalifying Leads 70
7.3 The Estimator's Team 74

Chapter 8

8.1 Complexity of the Materials Estimate 76
8.2 Where to Get the Numbers 77
8.3 Controlling and Tracking Materials Costs 78
8.4 Sample Lumber Chart 80
8.5 Contract Size 80
8.6 Market Analysis 81
8.7 Escalation Clause for Specified Building Materials 82
8.8 Substitution of Specified Materials 83

Chapter 9

9.1 Production Tracking 86
9.2 Daily Job Log 88
9.3 Payroll Data Entry 90
9.4 Change Order Form 91

Chapter 10

10.1 Sample of Estimate Versus Actual Expenses Report 94
10.2 Cash Flow Model 95

Chapter 11

11.1 Work Flow with Risk-to-Profit Points Identified 100

Chapter 12

12.1 Work Flow with Risk-to-Profit Points Identified 116

Acknowledgments

I appreciate the guidance of Senior Editor Doris Tennyson, a dedicated NAHB staff professional, whose career has made the lives of builders and remodelers more fulfilling through her work collecting information for the industry and disseminating it as BuilderBooks.

My thanks go to Marcia Asdal for proofing every page and lending constructive input to this work. I appreciate my daughters Ashley, Lindsey, Annie Rose, Kirsten, and Charlotte, whose prodding questioning, "How's the book coming?" spurred me to completion.

My thanks to Joan Brooks who builds these estimating defenses day in and day out for Asdal Builders in hopes of keeping us out of trouble. Her care and dedication to our core business allows me the freedom to cruise to the uncharted corners of the building industry.

I join BuilderBooks in thanking the following people for reviewing the manuscript and/or the outline: Alan Hanbury, Jr., CGR, CAPS, Newington, Connecticut; Bryan Patchan, Executive Officer, Frederick County Builders Association, Frederick, Maryland; Douglas L. Sutton, CGR, CAPS, Sutton Sliding & Remodeling, Springfield, Illinois; Michael Turner, CGR, The Home Service Store, Alpharetta, Georgia; and Robert Deppe, Robert Deppe Inc., Caledonia, Michigan.

Last, this book would not be possible without a willing audience who strive to build and rebuild America as professionals in an earnest quest to better serve America's home owners.

About the Author

Bill Asdal, owner of Asdal Builders, LLC, a remodeling and building firm in Chester, New Jersey, began building in 1973. In actively working for the industry, Bill has been chairman of the NAHB Remodelors™ Council, which represents the remodeling interests of the 225,000-member organization. The Council named Bill "Remodeler of the Year" in 2000. He is a past president of the Community Builders Association of New Jersey. Bill recently received an Achievement Award for "Educating an Industry" from Reed Business Information Systems. He is a frequent speaker on building and remodeling issues.

Bill is the co-author, with Wendy A. Jordan, of *The Paper Trail: Systems and Forms for a Well-Run Remodeling Company*. He was a driving force behind the creation of the award-winning *Professional Remodeler* magazine. He is a board member of the Energy and Environmental Building Association (www.EEBA.org) and a member of the Partnership for Advancing Technology Housing (www.PATHnet.org).

He recently completed a historic restoration, which is operating as The Raritan Inn Bed and Breakfast (www.raritaninn.com) and is underway with an age-restricted housing project in his hometown. Bill holds a Bachelor of Arts in Industrial Education and a Masters in Administration and Supervision. A former teacher and a licensed secondary school principal, he lives in northwestern New Jersey with his wife Marcia and five daughters.

Foreword

The book you are holding should be required reading for all builders and remodelers. In *Defensive Estimating: Protecting Your Profits*, Bill Asdal has done a masterful job of identifying the barriers to profit and answering the question, "How much should I charge?"

I would like to have a nickel for every time I've heard a builder tell the story of how he went to closing and the real estate agent walked out with a bigger check than he did. The lessons learned at the construction school of hard knocks are not quickly forgotten. Most builders and remodelers would prefer to learn from other people's mistakes, rather than experiencing them on their own.

Bill Asdal provides over 30 years of building experience in 12 easy to understand chapters. *Defensive Estimating* helps you figure out why you are in business and what you should be getting out of your business. He gives you systems to improve the process and ways to prevent you from chasing your tail trying to make ends meet. Within the pages of *Defensive Estimating*, you will find hundreds of ideas and simple suggestions that you will want to refer to time after time. Asdal takes the magic and science of estimating and turns it into an art. He gives you honest answers to tough questions. If you are serious in your commitment to improve your business and your profit, read this book!

—Sam Bradley
Sam Bradley Homes
Springfield, Missouri

Preface

"Who pays more for his shoes: a rich man or a poor man?" my mother asked me early in grade school. I fell for the simple answer that the rich man paid more, but she enlightened me. The rich man pays less because his shoes last far longer, and therefore, his annual cost is less than the poor man's. Indeed the poor man pays more over time. So the lesson was taught—but maybe only crystallized later—that life-cycle cost is far more important than first cost. Others followed this lesson by teaching me through examples of hard work, watching for dangerous risks, and always keeping a few safe paths out of nearly any situation. A long string of values and strategies were imparted to me, such as each of us might glean from caring parents.

This book is about such a lesson. A lesson in looking at your goals and creating a company that helps you fulfill them. It is about creating profits that cannot be endangered by carelessness or short vision so that each year is profitable. It teaches you to defend each estimate line item so that your planned profits are consistent and bankable. This book is directed to the creation of a defensive state of mind, so that it may become a system for generating and protecting profits.

You should be looking inward at your estimates and trying to find latent risks. Once you discover them, you need to defend against the risks to profit with clauses, terms, conditions, and disclaimers. We will review some of these risk-mitigating strategies in this text. You should not look for magic estimating formulas because they are not here.

As a society, we often read media coverage abhorring the profits of business. Excess profits are deemed a bad thing for society: the oil companies make too much money or corporate leadership makes too much salary. I suggest that this negative spin is wrong and that solid profits are a healthy way to keep a business and industry vibrant. Whether the stock holders benefit through dividends or a small owner-operator has a good year and banks some savings, the pursuit and achievement of profit is the reason to be in business. In defending the quest for profits through estimating, we secure our companies to face another day. A fellow builder shared with me the words of his immigrant grandfather, who entered the building business in America 90 years ago. He passed along the wisdom to know every dirty trick in the book. Not that you should use them, but to be sure they are not used against you in the building business. This company is now in its fourth generation.

My goal is for readers to rethink their quest for profits managed by "industry standards" and instead to tie their quests to their own needs in terms of financial security and personal growth. We each have a tolerance for risk in business, and you each need proportionate defenses to mitigate the risks you can't tolerate.

Our society speaks little about risk management for small businesses or individuals. Likewise, we give little effort to teaching decision making, financial management, personal skills development, process control, and a number of common sense topics. This book is a small step toward defensive thinking that may carry your business forward and just maybe protect some profits. The skills of defensive estimating readily transfer to other applications in life and other businesses. Given the potential volatility of our industry, some defensive thinking may be overdue.

CHAPTER 1

Understanding the Critical Number in an Estimate—Profit

Although every business has some commonalities—a registered name, a checking account, contracts, and deposit slips—the likenesses fade with the personalities and skill sets of the owners. Each small business reflects the skills and personalities of the owner. Some of these business-focused entities will flourish and generate a lifetime of income for their owners and employees. Other business owners will toil job to job for decades never quite knowing why they are not doing better. What is the difference between the two outcomes? Some owners run a business and have or buy technical knowledge for field work. Others have deep and perfected technical knowledge and find fulfillment in the production of work. A blend of both business and technical skills is likely to provide a productive balance that leads to stable companies.

Defensive Estimating directs its content to both groups. The first group will renew their vigilance in improving their bottom lines, and the second group will be reminded that a company is constructed of business systems, not simply a series of jobs. Builders, remodelers, and trade contractors already directing a successful and profitable business will find value in this book as they recommit to keeping a profit-centered outlook.

Throughout the book, the terms *builder*, *remodeler*, or *trade contractor* are generally interchangeable. Business success in general depends on planning for profit and defending its creation. To that end, any business organization could use the concepts in this book by substituting their attributes for the general concepts promoted here. The local retailer, grocer, restaurateur, or newspaper publisher—any of these people could refocus their firms away from gross sales as a measurement of success to the production and defense of net profits.

Defensive Estimating provides a profit-centered way to think about estimating. It presents a series of techniques that will sharpen your defensive skills in the quest to protect profits. These techniques will help you assemble pricing and develop personal skills in a way that defends your projections of profit. This book is not an estimating database with prices for thousands of items. It is not a software program into which you punch a few numbers and generate a plentiful array of proposal language and a magically created bottom line figure to present to a customer.

When your estimates are not focused on a contribution to the annual budget (through protection of profit), by default, the estimate becomes a step in the quest to acquire work. Work alone is not a satisfactory goal for a company. Work can be unprofitable, cyclical, and risky. Profit on the other hand is bankable and, if managed conservatively, secure.

The planning process for the creation of profits should reflect the companies' need to generate income. Chapter 3, "Establishing the Profit Number," will cover this topic in depth. Some companies are profitable because of the intellectual strength of the management team, and in others profit results from the hard work of the field team. Consistent profits should result from a solid plan that is well-executed at every level (Figure 1.1). Profit should be the reward for good planning and tireless management. Profits can vaporize with a contract error, an estimating mistake, or an unforeseen (and unprotected) action of a third party.

Work Flow

A clear work flow in estimating is repetitive and logical. The basic steps shown in the diagram in Figure 1.2 are data collection, assembly and estimate, customer approval, data handoff to production, review estimate versus actual costs, and update your pricing database.

FIGURE 1.1 Title Page of Business Plan

Your Building Company

February 20??

Business Plan
Copy Number 1

This document contains confidential and proprietary information belonging exclusively to Your Building Company, LLC.

Prepared for:
Your Name
(Managing Member)
Your Address
Your Phone

Prepared by Your Team

This is a business plan. It does not imply an offering of securities.

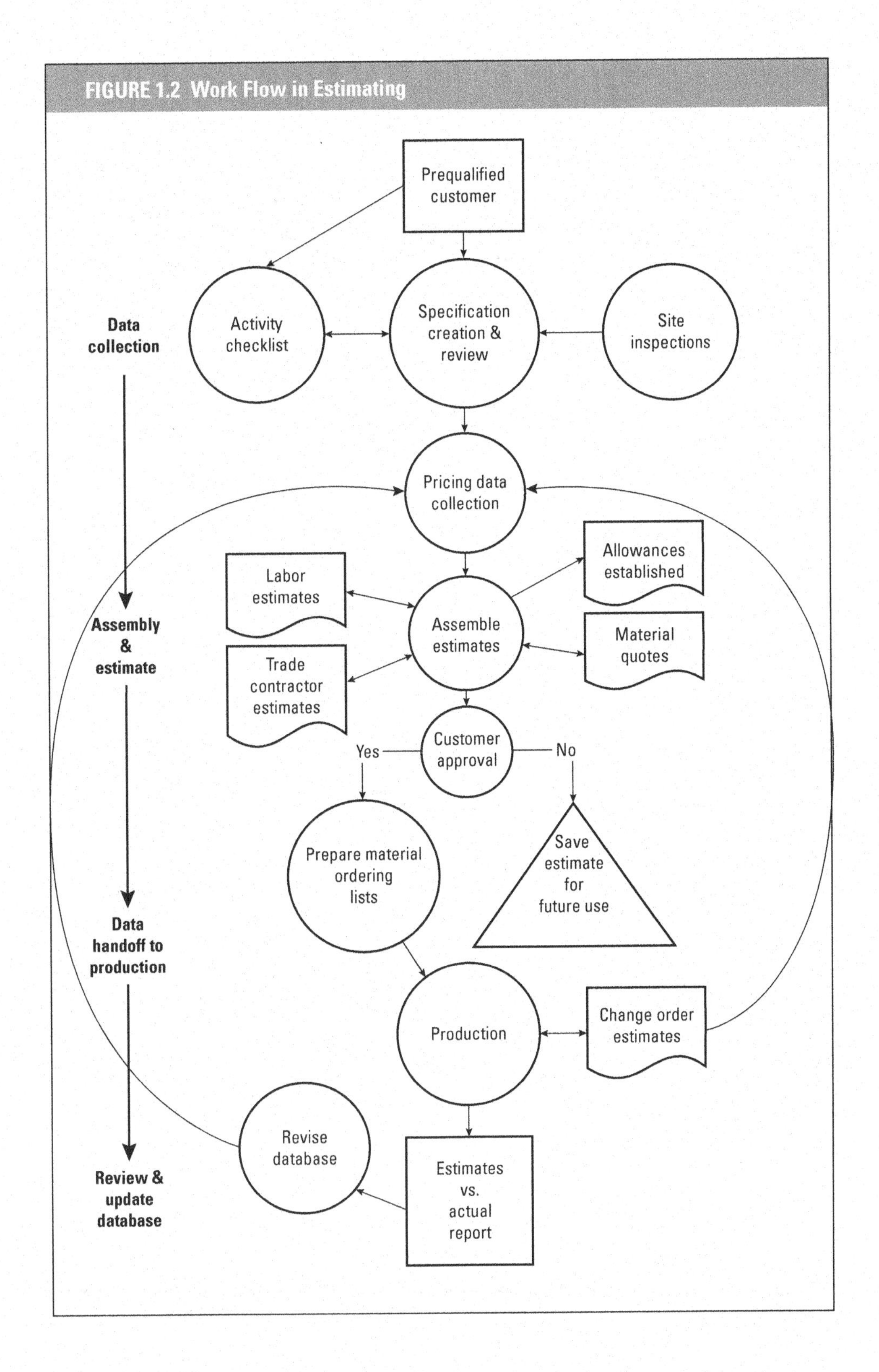

Being able to look at estimating as a process is a fundamental step in mastering the skill of estimating. Getting the numbers of "sticks and bricks" only can be devastating to profits because it is neither a process nor inclusive of all the activities required to build a job. The estimating work flow in Figure 1.2 is a process that repeats for every job. If the builder or remodeler estimated the costs and missed the activity of data collection from a site visit, the omissions could be devastating. Some builders and remodelers may try to put numbers to paper without collecting all the specifications. This practice may be fine for an early budget discussion, but it will hardly suffice for an outline of a profitable job. *Systems thinking* is the format that can take any business to a heightened awareness of processes. If you identify the repetitive steps while you constantly improve the process, the likelihood of consistent results in business goes up.

In my view, estimating is not a math puzzle with a correct answer.

Plenty of jobs are sold without fully defining the scope of work. Unless a change order covers the difference, the builder or remodeler later has to pay with the profit line item for a customer's delayed clarification of what the customer thought he or she wanted. A defenseless line item puts profits in danger. This workflow starts with a prequalified customer. Chapter 7, "Minimizing the Workload," explores how to do fewer estimates and still remain fully engaged and profitable.

Everyone doing estimates has a process. Once you outline the steps in your process, you will probably

- improve the workflow by eliminating redundant steps
- reassign portions of the information-gathering to others
- use the freshest technology to gather data and put it in place

You can now professionally assemble the estimate. Steps in the process are clear, the methodology logical, and the outcomes well defined.

Estimating is a game of business activity. Profit is the king of the board, and you must protect it. The contents of this entire book focus on the creation and defense of profits. The concepts should help you rethink how you approach estimating. The strategies should shape your customer and supplier relationships. Chapters 8–10 cover materials tracking, production tracking, and financial analysis as they relate to estimating. A number of contract strategies in Chapter 11 defend the estimated profit from the signed contract throughout the rest of the job. Finally, the profitable results of a logical and well-executed estimate for a building or remodeling job can change your life because a continuum of positive cash flow can come your way (Figure 1.3).

Defining a Professional

A professional is someone who works on his or her business, not just in it. Many tradespeople embark upon getting their own work and having their names on their trucks. Yes, they each own a company, but they may not own a business. A company is merely a legal entity, and ownership can be as simple as registering a trade name and paying a small registration fee. However, a business implies a series of processes leading to the creation of profits. A business needs customers, marketing to find those customers, sales, and a product or services. It strives to improve systems, minimize costs, and

FIGURE 1.3 A Profitable Day Building to Net Worth

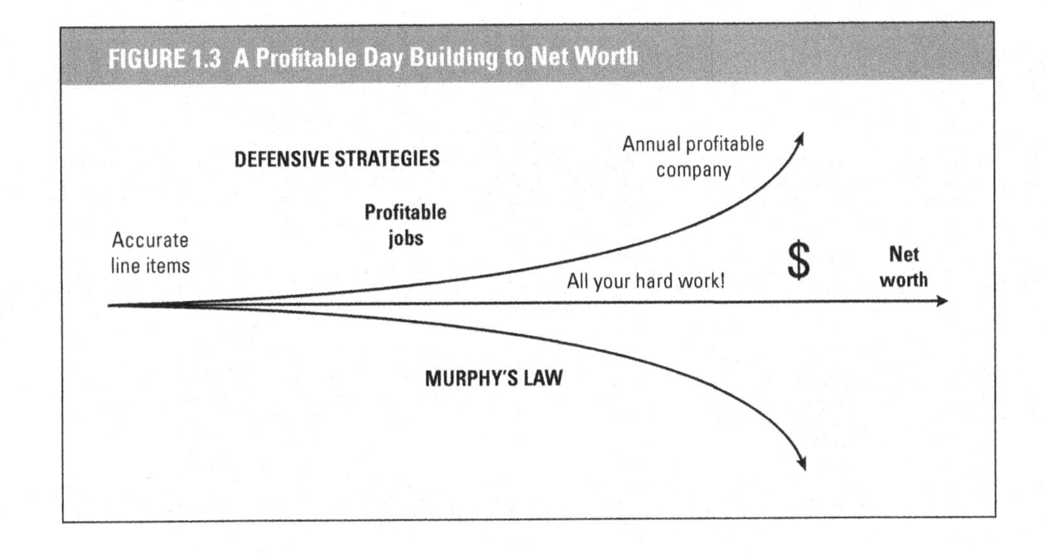

A business should support the needs and lifestyles of those who make it successful.

maximize profits. Management of these processes is the key to making a company into a business. An owner must work on the business or hire someone who will. Having a lifetime of work without working on the business is like having a job without a boss. Would you really want to work for a company

- that has nobody working on business processes?
- in which nobody is working on evaluation or measurement of activities?
- in which nobody is working on getting the most out of the company?
- with no boss?

No one should, but many do.

Establishing the Overview

Where is your business taking you?

A solid business should enable the owner and employees to achieve any number of life's goals (Figure 1.4). A more common perspective makes the end goal the act of business ownership alone. Infomercials often harp on this idea: "Own your own business!" This common perspective can surely limit your experiences in life and lead to a fixation on the business without personal growth. To the extent possible, you should form a business to help you achieve your hopes and dreams for your life (Figure 1.5). The benefit of building a company to support a lifestyle is that you can specify the degree of each that you can manage. A lifestyle and business can be mutually supportive. In order for them to do so, however, you need to connect the dots of your goals and how your daily activities support achieving your goals. You will see the estimating connection shortly in Chapter 2, "Establishing the Profit Number."

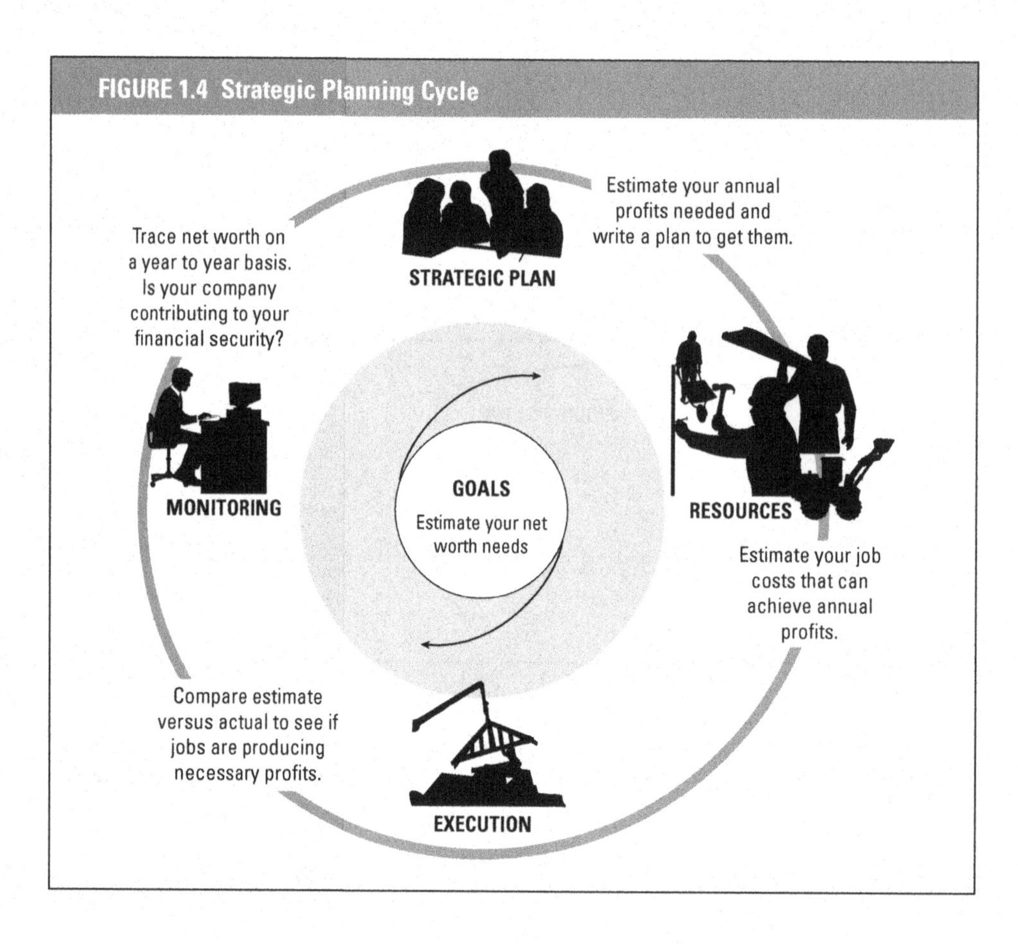
FIGURE 1.4 Strategic Planning Cycle
STRATEGIC PLAN
Estimate your annual profits needed and write a plan to get them.
RESOURCES
Estimate your job costs that can achieve annual profits.
EXECUTION
Compare estimate versus actual to see if jobs are producing necessary profits.
MONITORING
Trace net worth on a year to year basis. Is your company contributing to your financial security?
GOALS
Estimate your net worth needs

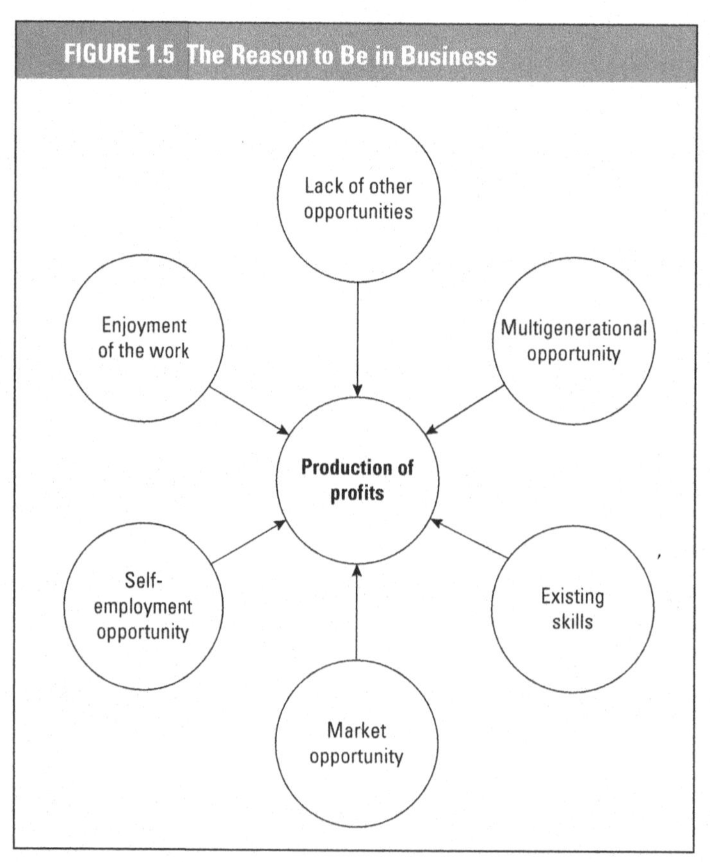
FIGURE 1.5 The Reason to Be in Business
Lack of other opportunities
Multigenerational opportunity
Existing skills
Market opportunity
Self-employment opportunity
Enjoyment of the work
Production of profits

If a company's role is to support a lifestyle, you must define this lifestyle and how much you will need to pay for it (Figure 1.6). A company may be a stand-alone legal entity, but its burden is to return income to its owner or shareholders as profits.

Building a company that fulfills your life's expectations is indeed empowering. Your company must rely on the production of profits to fund a personal budget and lifestyle.

FIGURE 1.6 A Remodeling Company Supports a Change in Lifestyle

Renewed Purpose

Craig A. Shutt

Eric Schneider spent 18 years building his remodeling business by doing exceptional work, winning awards, building a reputation and generating lots of publicity. The effort paid off as the company grew to a $2.5 million business. And then early last year, he left. Through a transitional process that required the establishment of several new systems, he turned over control of his business to his former accountant and didn't look back. Today, he receives weekly updates and salary but offers only advice on how to run the business.

"I'm 52 years old, and I didn't want to wind up being 65 and feeling that I missed an opportunity to do something else with my life, " says Schneider, the new chairman of the board of Eren Design and Construction, Inc., in Tucson, Arizona. "I wanted to do other things as well as run my business, and I found a way to stop that I could exit from it without causing it to stop running well."

The change wasn't made overnight, and it wasn't done without making significant alterations to the business. But achieving it wasn't as difficult as most remodelers might think. "What we did, anybody can do," says the new chief executive officer, Janice Donald. "When an owner's passion is no longer expressed by running the business, it might be time to move on. For some, it would be easy to do and for others it would take more effort. But for all, it might make good business sense."

The concept first presented itself to Schneider in 1998 when he won a National Remodeling Quality Gold Award. One of the speakers at the event, Silver Award winner Bill Asdal, CGR, of Asdal Builders in Chester, NJ, discussed using his business to leverage other financial and personal interests. He stressed that remodelers should know why they're in business and what the purpose of their business is.

"It was the most empowering thing I'd ever heard," says Schneider. "I wondered if I could do that, and if I stepped away, if my business would be able to run itself. Little did you realize that I'd already inadvertently set up my business to make that possible."

From Shutt, Craig A. "Renewed Purpose." *Professional Remodeler.* February 2001, p. 94–96. Reprinted with permission.

CHAPTER 2

Establish the Company Profit Number Based on Your Income Needs

The company profit number should not be a vaporous goal of "I want to make all I can!" This goal is not attainable because it includes no number. You would simply never get there. The number you set out to earn should be indexed to what it costs you to live. Surely you need to earn that amount for sustenance and 50% more for savings, growth, market fluctuations, and other contingencies.

Establish Your Income Needs: Set a Personal Budget

When a speaker asks seminar attendees, "How many of you have a personal budget?" the response is regularly less than 10%. The number of attendees who confirm the existence of a company budget is only slightly higher. Creating and managing your personal finances is a significant personal step. If you can't quantify, manage, and be accountable for yourself financially, you are unlikely to do so for a vaporous entity such as your company. Tying the company performance to the fulfillment of personal financial needs is critical to overall success.

A planning exercise of quantifying personal needs provides the foundation for driving the company you develop to support your needs. Initially, entrepreneurial builders and remodelers will often state that they want to make "as much as they can" from their work. Others might describe a lifestyle without establishing those lifestyle costs. Also, some builders and remodelers might describe their interest in creating financial security without specifying a cost for that desired level of security. All these answers beg the question, "How much is that?" The answer lies in your ability to estimate for personal consumption and then live within that number. This estimating challenge resembles creating a job cost estimate and delivering it on budget, so a personal budget also is a target for spending that you need to adhere to with dedication.

Estimating is a process, and it works well beyond construction costs. Without estimating and managing your personal expenses, to index your company's income to your personal expenses would be impossible. These two items are related and need to be managed to keep the relationship proportionate. Generating a company budget that produces a minimum of 150% of the owner's personal expense needs allows a 50% excess for personal and business growth or the security of increased savings to enhance your net worth. If you look at net income before owner's compensation, whether draws, salary, hourly, or commission, you have a number you can use and compare with any company regardless of whether the firm is an S or C corporation, a limited liability corporation (LLC), a partnership, or a sole proprietor. The company doesn't pay the owner's personal expenses, but it must provide for them and have net profit of its own to fuel growth and protect against rainy days.

I have taught dozens of seminars on this topic, and when I ask the question of attendees, "How many of you have a personal budget?" The response is regularly less than 10%.

If you maintain a firm grip on personal expenses and a have a proven track record of staying on budget, please skip the balance of this foundational chapter and move to Chapter 3, "Competencies," and beyond to build estimates that produce profits that fulfill your personal needs. If you are curious or do not have a personal budget within your grasp, hang on for a few more pages. I will review the basics and set the stage for a relationship between income from your company and personal expenses.

Some discussion takes place in our collective youth about personal budgeting. Whether at home, at school, or through self-discovery, learning about the balance of income and expenses takes place. The United States bankruptcy rate is currently at about 0.5% of the population per year. This figure shows that not everyone has learned the lesson or has been able to manage the risks. The U.S. Department of Commerce calculates the national savings rate to be 0.2%. This percentage provides further evidence that more attention could be paid to balancing income and expenses. Many retirees are counting heavily on social security benefits to supplement retirement because they may not have saved enough along the way. Your company needs to not only pay the current expenses for the firm, but it also needs to generate enough money to pay your personal expenses and contribute to your savings.

The income from a building or remodeling business is *active income*, income that you derive from the labors of your work. Along the way you should use this *active income* to pursue *passive income*, income you derive from your investments in the labor of others. Passive income can come from stocks, bonds, real estate, or any other investment that can produce a continual stream of ongoing funds. Creating passive income should be a business proposition in itself.

The core work of business is initially to create active income, stabilize its flow, and minimize the risk to its ongoing production of future dollars. As a business (and its owner) matures, the owner must spend some time creating passive income lest the owner toil away at churning income against expenses without creating the financial security of a passive income stream. Passive income has a series of advantages over active income:

- It is not subject to self-employment tax.
- It can be tax exempt if you invest it in tax-free instruments.
- Equity is not subject to tax.
- Income increases annually through compounding or escalation.

- Payment of mortgage principal provides forced savings.
- Asset value rises with inflation.

Active income, on which business owners consistently focus, has a different set of attributes. A blend of both endeavors is indeed desirable. Active income is

- cyclical
- fully taxable
- subject to self-employment tax
- subject to income tax

The chart in Figure 2.1 plots the passive income line against active income and personal expenses. You not only, especially need to understand the personal expenses line, but you also need to control it over time. Surely spikes in spending occur for needs for health reasons or college expenses, but you can anticipate and plan for some of these through specific savings or insurance plans. The active income line results from your building or remodeling endeavors and, like expenses, you need to manage it for incremental growth. You should be able to project some increases in active income and work toward their realization. The line for income is intentionally plotted with a slightly higher rate (5%) of increase than the expenses line (4%). This small divergence over time has a big impact on the compounding of net worth and generation of passive income.

FIGURE 2.1 Passive Versus Active Income

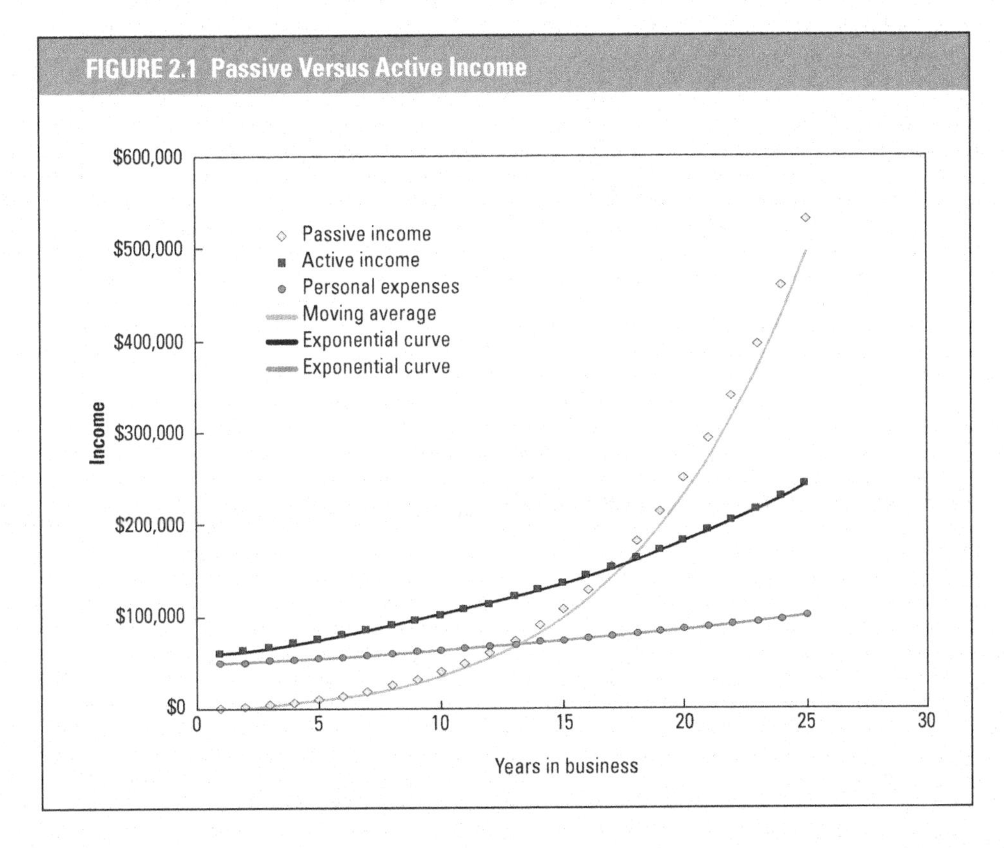

The actual numbers in the chart in Figure 2.1 are meant only for illustrative purposes. What is interesting in these growth curves is that the active income and expense lines are generally incremental or linear. The passive income line is a parabola, or curve, because of compounding and divergence of the income and the expense lines. These lines can be bent sharply depending upon the growth in your active business or the constraints on spending.

The difference in income versus spending is the only modifier of the passive curve initially. The amount of risk and rate of return in passive income is the second modifier. A high-risk passive investment compounded over time surely has a dramatically, differently shaped curve for passive income. In the later years, the spread between active income and expenses gets larger and allows for a faster investment in a passive stream. The passive income curve accommodates the additional savings and also the compounding of the underlying assets. The one magical point on this chart is where the passive income curve intersects with the personal expenses line; this juncture is the turning point in financial freedom (Figure 2.2).

Moving forward over time, this circled point shows sufficient passive income to support the planned personal expenses. After this point, you can soften the active income line by not working as many hours or by committing time to other activities while still meeting your basic needs. Financial freedom

Expressed as a percentage, I stated earlier that you need 150% of your personal expenses covered by the active business income. The next step is to achieve 150% of those personal expenses through passive income.

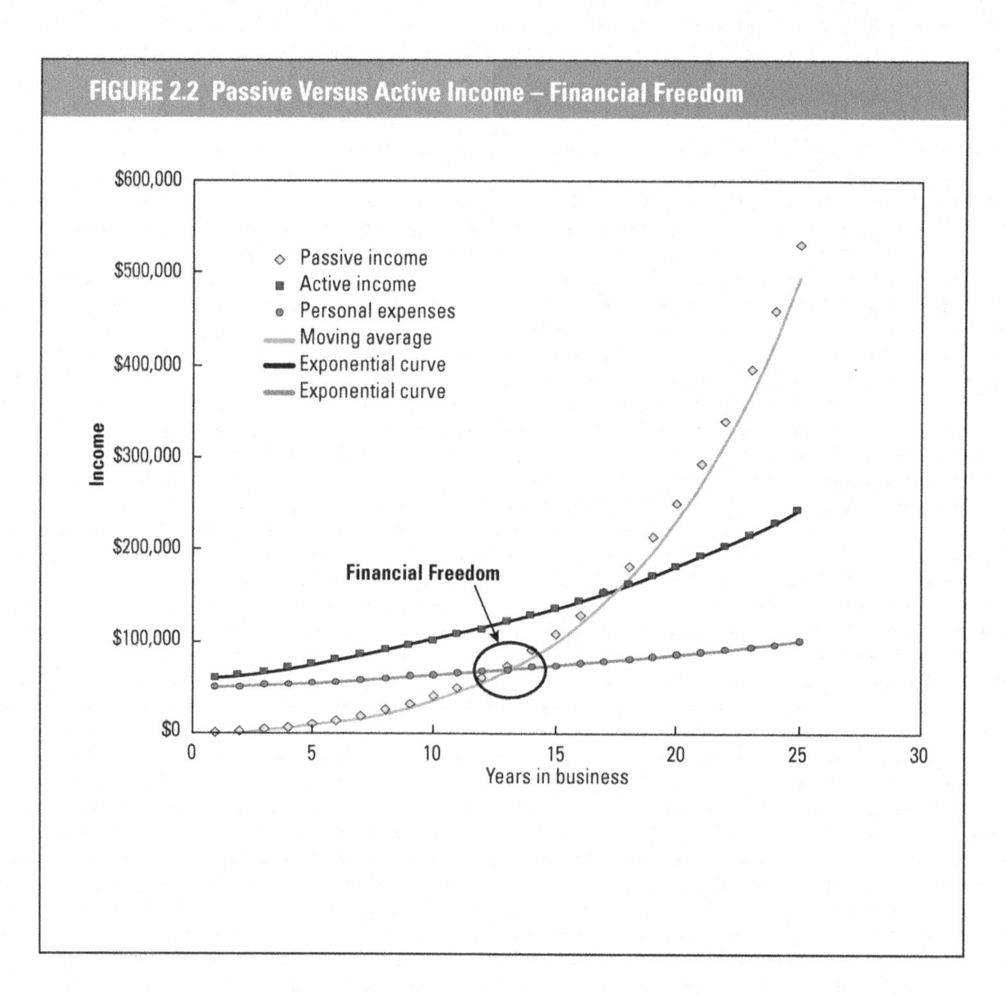

FIGURE 2.2 Passive Versus Active Income – Financial Freedom

can have a life-changing impact on a small-volume business owner who has managed with active income for years.

The parabola shape softens rapidly if active income work ceases. The curve gets sharper going forward as you devote additional active income to creating passive income. In addition to the simple reason of continued love of the work, staying involved with active income production for years beyond the point of financial security can have a positive impact on the generation of additional passive income.

You have several ways to measure your success at creating this passive income stream. The simplest is to look at a personal balance sheet each year and compare your net worth on a year-over-year basis. This measuring tool challenges you to make a serious commitment to securing your future through savings investment. Some aggressive investing may be able to push your net worth to grow in the range of 10% a year. A second way to measure your success at creating passive income is to compare an index of your personal spending to your passive income stream.

As you can see from the curve of the parabola in Figure 2.2, in just a few more years at this formula, increases of 200%, 300%, and more are possible when you continue the hard work of creating active income and suppressing expenses. The higher the ratio of passive income to expenses, the more security you layer on your financial life.

Banks demand personal financial statements as a requirement for any loan. You need to understand the net number of this personal planning and evaluation exercise to be sure your estimates generate sufficient income to pay your bills and generate savings. Profitable line items on an estimate lead to a series of profitable jobs. These jobs, in turn, produce a profitable year. A series of well-managed profitable years can generate a secure net worth for long-term financial security. Every line item in an estimate contributes to gross profits. Each item needs to be mathematically correct, but it also adds a level of responsibility for overall management of money.

The oversimplified chart of personal expenses in Figure 2.3 is only for illustrative purposes. Any bank, financial software package, or quick Web search will yield a far more complete document for your use in planning. In short, you need to account for all your expenses including contingencies and frivolous expenses that are sure to occur. Putting a dent in a budget at home through mismanagement is no more tolerable that dinging your company budget with uncontrolled expenses or a failure to generate income.

One possibility is generating a company budget that produces a minimum of 150% of the owner's personal expenses provides a 50%

FIGURE 2.3 Personal Financial Statement

Personal Income	
Schedule C	
W-2	
Interest	
TOTAL INCOME	
Personal Expenses	
Household	
Mortgage and debt service	
Vacation	
Personal	
TOTAL EXPENSES	
Net Income	

excess for personal and business growth or for the security of increased savings that enhance your net worth. The next few pages will address a concept of risk to companies introduced in this graphic. The relevant lines in an earlier chart (Figure 2.2) were the (active) income line and the (personal) expenses line. Now, a smart exercise would be to fill in a sample budget with the help and consensus of anyone else who is spending the family funds. The number you decide upon for personal expenses will serve as the basis for creating the company budget that fulfills this expectation.

One person can "bust" a budget as easily as two; consensus and cooperation are the watchwords in keeping the budget in balance.

Establish Your Savings Needs

You can readily discern the rate of your need for savings from some simple financial modeling. The variables include any dependents you may have, your spending needs, the rate of return, the rate of inflation, the time until you will start withdrawing the funds, and the length of time the money must be available for your use. Any number of Web sites offer interactive calculators that will determine how much savings you need to create a pool of money large enough to sustain a family without active income after retirement. Figuring this amount into each year's budget for personal savings is a critical step.

Equally imperative, you need to make sure the company funds this amount from a series of profitable jobs (Figure 2.4). If you can save $10,000

FIGURE 2.4 Gross Sales Planned by Income Needs

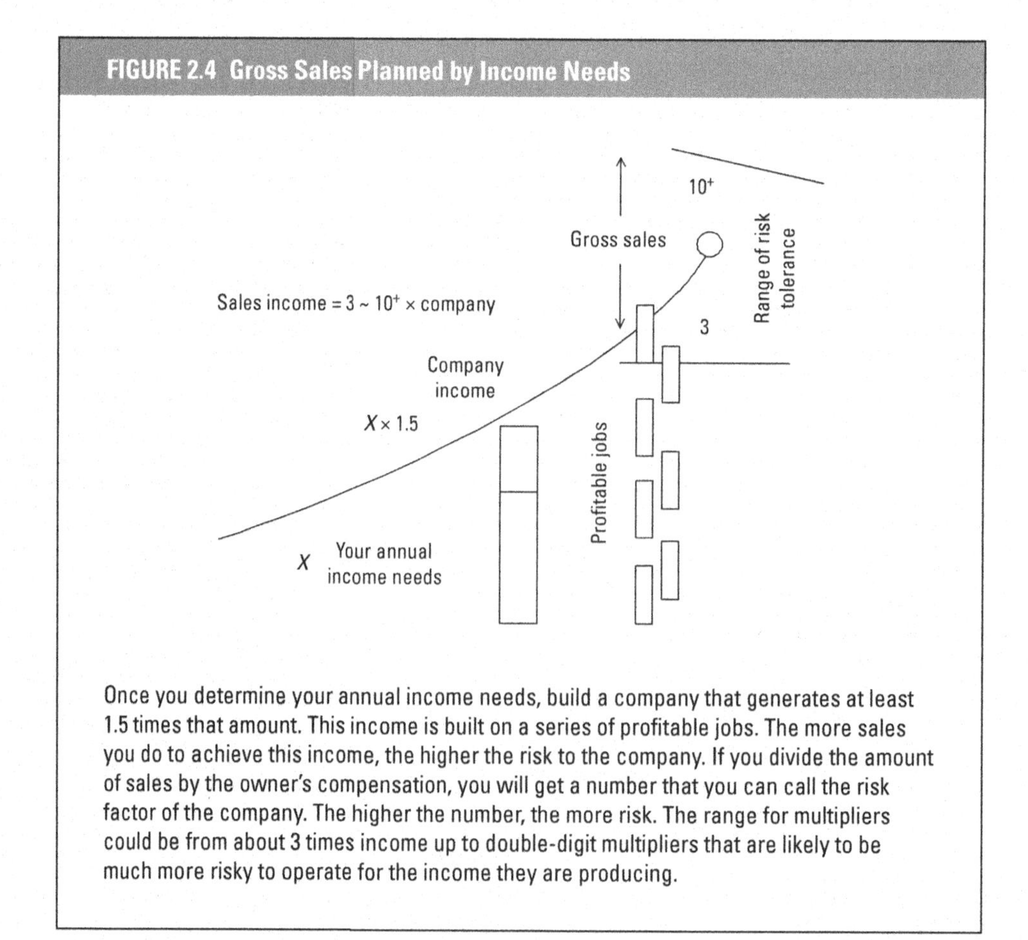

Once you determine your annual income needs, build a company that generates at least 1.5 times that amount. This income is built on a series of profitable jobs. The more sales you do to achieve this income, the higher the risk to the company. If you divide the amount of sales by the owner's compensation, you will get a number that you can call the risk factor of the company. The higher the number, the more risk. The range for multipliers could be from about 3 times income up to double-digit multipliers that are likely to be much more risky to operate for the income they are producing.

in a year and increase that annual savings rate, as well as average the rate of return on your investments at 10% per year, you would have over $500,000 in 15 years. This number doubles in the next 5 years to over a million dollars, and doubles again in the next 5 years to over $2 million. Profits turned into savings that you can effectively manage are powerful indeed.

Create a Company Budget to Meet Your Personal Needs

You can express profit as a percentage of gross sales, the traditional way to look at the ratio. The rule of thumb in the remodeling industry has been that 33%–40% of gross sales should be gross profit and the owner should be taking 20% of gross sales in combined salary and net profits. The *NAHB Remodeler's Cost of Doing Business Study* shows more conservative numbers being put on the books.[1]

Another way to look at this correlation is to state your intended earnings and then assign a scaled risk factor to intended earnings to indicate how much risk you are willing to take to create your income. You can determine your risk factor by dividing gross sales by your income. A builder doing $3,000,000 in gross sales with an income of $150,000 has a risk factor of 20. A remodeler doing $600,000 in gross sales with an income of $150,000 has a risk factor of 4. The implication here is that the lower the risk factor the more likely the company is to be stable over time. So after a bit of introspection on your personal risk tolerance and a thoughtful look at your personal financial needs, you should be able to produce a projected gross sales number for a company that accomplishes your financial needs within your comfort zone. Put into a formula, this method would read:

R (risk tolerance factor) × I (income needs) = P (projected gross sales).

No pun is intended by the fact that this formula for business reads RIP. Why anyone would want more risk may not be readily evident. As the risk goes up, so does the opportunity for higher returns. When the gross sales are high, if you can maintain the return-to-owner ratios, you can bank some serious money. These spikes of income (or sustainable increased active income) can move the passive income parabola sharply higher and, thus, ultimately reduce overall risk as the passive income stream gets larger.

A second way to do a quick estimate for an annual budget for an existing company is to use last years' numbers as a baseline. If you are uncomfortable with your income history, repairing that history will be far easier at the budget level. Evaluating the pitfalls of the past and then managing your way to more acceptable profits is easier than simply creating more sales that continue to yield unacceptable profits.

Initial population of a budget can come from industry standards or your company history. These estimates are only for the company, and you and your staff should treat them as such. You should measure and manage your business to achieve this budget during the year. A sample budget for a small-volume remodeling firm appears in Figure 2.5.

The adage I heard as a kid was, "It takes money to make money." I also think that you have to take some risks now to lower later risks. Managing the possible pitfalls is the role of a good manager.

FIGURE 2.5 Sample Budget for a Small-Volume Remodeling Firm

Annual Budget	TOTAL Jan–Dec 08
Ordinary Income/Expense	
Income	
4000 · Gross Sales	744,000.00
Total Income	744,000.00
Cost of Goods Sold	
5000 · Job Related Costs	
5100 · Materials	258,000.00
5125 · Plans, Permits, Approvals, etc.	21,780.00
5150 · Trade Contractors	111,600.00
5200 · Job Related Labor Costs	
5210 · Job Labor (Gross Wages)	80,400.00
5230 · Employer Payroll Tax Costs	10,200.00
Total 5200 · Job Related Labor Costs	90,600.00
Total 5000 · Job Related Costs	481,980.00
Total COGS	481,980.00
Gross Profit	**262,020.00**
Expense	
6000 · Overhead Costs	
6010 · Advertising	3,281.76
6040 · Dues/Journals/Seminars	3,041.28
6045 · Donations	240.00
6050 · Entertainment-Food	3,000.00
6070 · Insurance	
6071 · Specialty Contractors Package	6,000.00
6072 · Worker's Comp	7,200.00
6073 · Other Auto	328.44
6074 · Fire Insurance–Liability	931.08
6075 · Health	12,000.00
Total 6070 · Insurance	26,459.52
6100 · Office Supplies	1,200.00
6120 · Office Salaries	20,400.00
6130 · Professional Fees	
6131 · Accounting	2,580.00
6132 · Legal Fees	6,000.00
Total 6130 · Professional Fees	8,580.00
6150 · Repairs & Maintenance	1,200.00
6160 · Service Charges	0.00
6185 · State Annual Report Fee	50.04
6190 · Telephone	9,312.00
6200 · Tools	1,200.00
6210 · Travel/Lodging	1,200.00
6220 · Utilities	3,202.68
6240 · Vehicle Expenses	
6241 · Gas & Oil	5,700.00
6242 · Repairs & Maintenance	4,200.00
6243 · Registration & License	840.00
Total 6240 · Vehicle Expenses	10,740.00
Total 6000 · Overhead Costs	93,107.28
Total Expense	93,107.28
Net Ordinary Income	158,912.72
Net Income	**158,912.72**

The amount of profit that each job produces needs to support the yearly budget. The yearly budget cycle has no magic other than for tax reporting and is little more than a window through which to view the health of your company. A year gives us a common denominator for comparison to past performance and to industry standards. If, in a year, you want to make $100,000, you can do so in a number of ways. The solution could be working one contract of $1 million with a 10% net payout or $2 million with a 5% net payout. As noted earlier, the difference is in the risk. Completing more small jobs to the extent that they do not raise overhead significantly may be the less risky strategy. Chapter 11, "Defending the Profit Line," outlines a number of ways to minimize risk to profits.

The next logical question would be how much the company has to make on a specific job to make an effective contribution to the company profits for the year. The math answer for this question is to mimic the percentage of yearly profits compared to gross sales and use this formula for every job. A direct proportion is the easy math answer.

The total profit generated on small jobs may not be enough in total dollars to make a significant contribution to the annual budget, so you are likely to need to increase the percentage that you must obtain for this work. The steps you take in the process—gathering the lead, selling the customer, estimating and contracting the work, completing inspections, and job closeout—are similar no matter what size the job is. And every job grabs some of the overhead expenses. These costs often burden the company with a minimum fixed expense, so if the sale price is lower, the percentage of markup has to be higher to accommodate your expenses. Rather than a fixed rate of markup on costs and overhead, this understanding leads to a scalable markup for profits.

I have often been asked if the profit percentage should be lower on a large job within your year of work. The answer is yes. The fact that is often forgotten is that the percentage of profit on smaller jobs needs to be higher.

A local remodeler who did an outstanding job of tracking costs would often report that his most profitable jobs, by percentage of revenue, were completed in one day. This profit resulted not from the rate of markup nor the speed of completion. If you are watching the expense percentages in the estimate, your overhead will likely drop in a large job because the steps and clarity requirements stay much the same. If small jobs require just as much hand-holding, they need to be marked up higher because the total profit dollars are smaller.

This concept could be called *scalable markup* (Figure 2.6). The mathematician would call it a logarithmic scale in which the intervals between numbers are not uniform. The hard dots connected by the straight lines represent the core numbers. The sweeping curve is the logarithmic trend line.

This chart reflects some projections for a company in the $2,000,000 gross sales range. Several divergent industry ratios could be mapped similarly. A grocery store owner may work on a net income 2%–3% of sales while making millions, and the hotdog vendor may mark up the $0.50 product by 400% to make $2 gross per sale. Markup alone is not the difference in the business models. Among other factors, risk tolerance plays a critical role in which company you build. Do you want to be the hotdog vendor or own the grocery store? Both options are feasible in the retail food business.

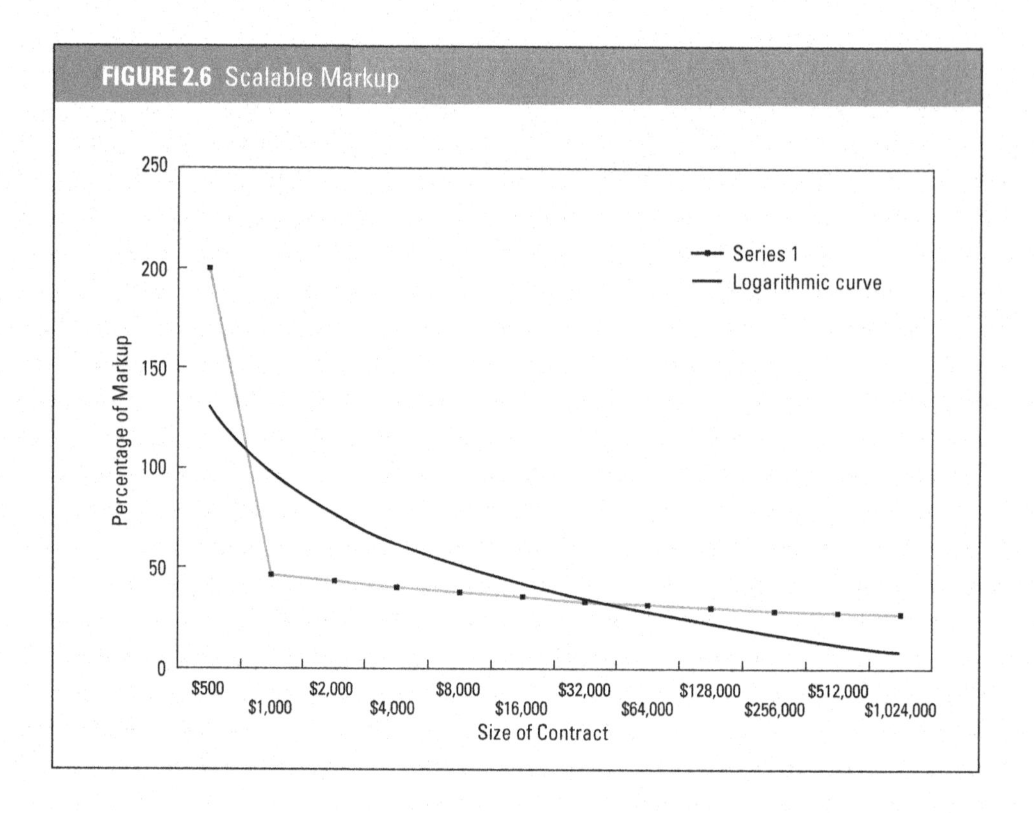

For decades in the building and remodeling industries, a running dialogue has considered what number is the "right" markup number. Consensus seems to point toward a high-end target of 67%, which renders a 40% gross profit. Each company's cost of doing business must be reflected in the calculations, and yours could be lower and should be the basis for your company's pricing decisions.

Using a standard 67%, can be a dangerous way to look at markup because your intent is to generate profits. Keeping your eye on your profits and adjusting the markup to create profit is a far more logical method of staying at the level of net income that you need. With a markup of 67% on a series of small-to-medium–size jobs and not enough gross sales for the year, you could work until you drop without ever achieving financial security or worse go bankrupt trying.

CHAPTER 3

Competencies for Building and Remodeling Success—Estimating

Builders and remodelers need six competencies to run a successful construction business. You must master and integrate all six of them into your business. Competency describes a set of acquired skills that you hone through education, on-the-job training, a firm grasp of core information, ongoing professional development, and cultivated personal skills that you recognize and control from within. Competency is an on-going quest for currency and mastery of a subject matter. Attending an estimating class does not confer competency. Learning a software program does not mean you have mastered estimating. A stagnant plateau of knowledge does not produce good estimates. You need skills and the dedication to keep them current.

The six competencies you need for success in the building and remodeling industries are

- construction communication
- marketing and sales
- estimating
- production and safety
- business management
- financial management

Shown as a DNA model that makes a building company succeed, Figure 3.1 illustrates just how fundamental these six competencies are to success. Deficiency or loss of any of the skills makes the risk of business failure rise dramatically. A company could not survive long in the marketplace missing any of these critical skill sets. The absence of one may not show up for years, but its ultimate unveiling can be fatal in the building business.

Construction Communication

The ability to communicate in three mediums is essential for contracting success. These mediums are written, oral, and graphic communications. Written communication includes the ability to craft specific, unambiguous specifications and contract terms and conditions. (The sophisticated English classes of today call the understanding of written documents *critical reading*.)

I have been called as an expert witness in numerous construction court cases, and I can confirm that parsing words from what you say to what a customer hears can involve a wide disconnect. (Addendum Clause No. 2 in Chapters 11 and 12 addresses this gap.)

FIGURE 3.1 How Estimating Fits into the Competencies for Overall Success

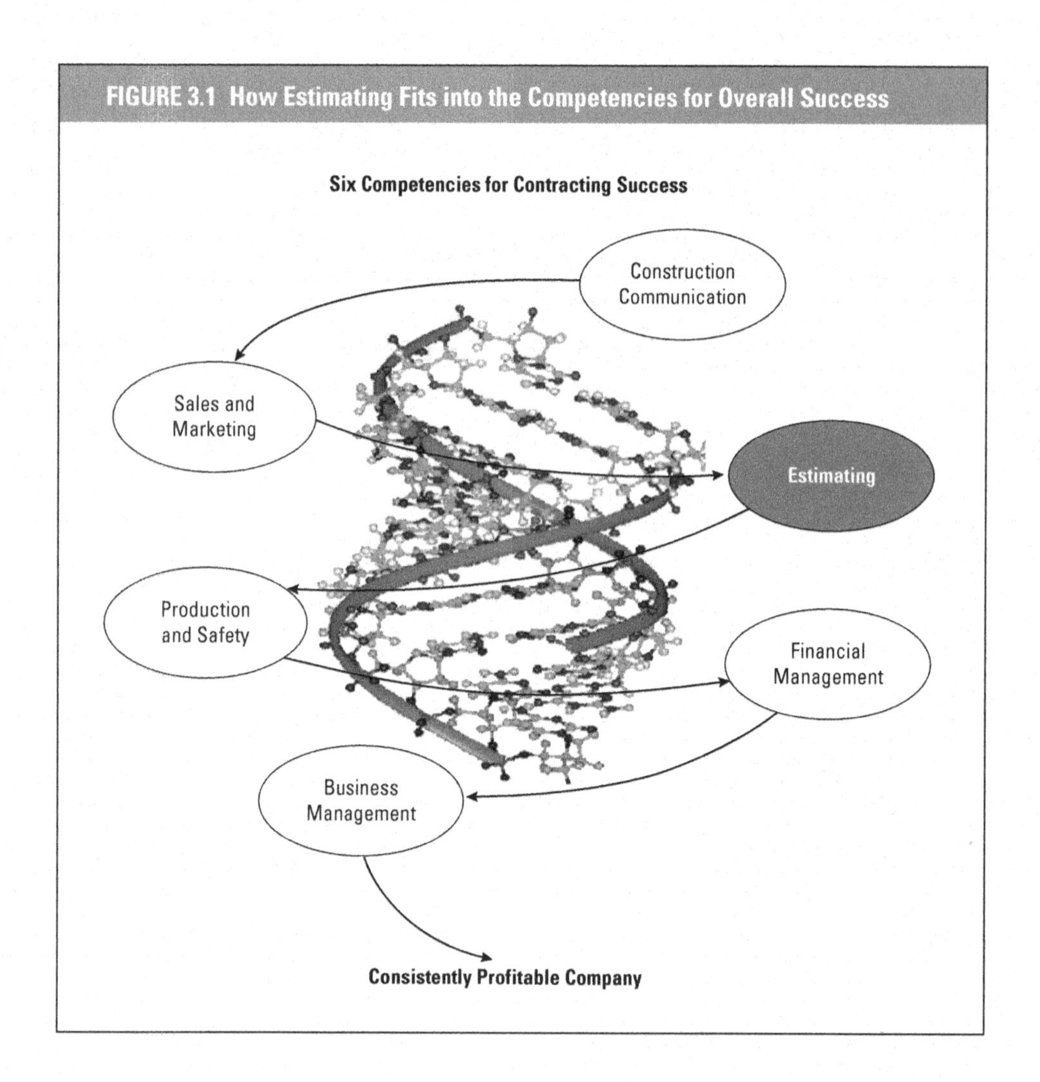

If you were to miss a specification detail in one of the written text sections on a plan, that detail would reduce your profit. The templated and often cut-and-paste specifications added to the last pages of a set of plans become a critical part of the document. A quick read looking for unusual language with a highlighter in hand regularly unearths some cost-inducing detail; you can then log these back onto the estimate sheet with a concurrent cost.

Oral communication includes both critical listening and crisp, clear speaking skills. After listening to a client describe what he or she wants, you should respond to words that trigger added cost by amending the scope of work, the estimated cost, or the language that defends the profit line. When describing what you will produce and when you will do the work, you need to be precise.

You probably have experienced a dominating speaker who employs endless words to make a point. These underskilled communicators may use 300 words to make a 30-word point. This technique demands the listener figure out which 30 of the 300 words are critical to the point of order. Far better, the speaker should filter his or her words to the point and communicate that idea in a clear and crisp delivery. This skill clearly is worth cultivating.

Graphic communication uses drawings to convey an idea. The draftperson's world includes sketches, working drawings, details, overlay plans for mechanicals, elevations, perspective drawings, and exploded sections—all meant to convey a clear vision. An estimator must have an expert's critical knowledge of what he or she is reading on a plan and the baseline skill to draw the work him- or herself.

In construction, any communication needs a minimum of two of these three mediums. An example is the assembly instructions for a barbeque grill with text to follow and usually also some exploded-view drawings to help the buyer complete assembly of the grill. Think about how many mediums you use to describe a detail of construction performance and make a conscious effort to have a minimum of two. The adage of "talking with your hands" is surely helpful in the creation of an "air graphic" or visualization.

Marketing and Sales

Through a screening process, you narrow your building or remodeling opportunities to the select few jobs that your company estimates and builds. Some builders/remodelers work from job to job without ever putting a process in place that winnows an undefined large mass of consumers to a finite number of actual estimates and a tighter-yet pool of contracts to be built.

Marketing your firm is a skill. Closing sales is a skill. However, they are not interchangeable though they are interdependent. To create a market image for your work, begin with the simple task of gathering references or referrals from current customers. A detailed guide to building these skills is fodder for another book, but without a market image and without sales you cannot sustain your company.

Estimating

Yes, this process-driven skill is the subject of this book. Get the estimates wrong or fail to protect profits, and the company will vanish.

Production and Safety

A traditional role in the development of production skills in building and remodeling begins by carrying lumber. Beginners further develop those skills as apprentices, carpenters, master carpenters, then as builders or remodelers, to developers.

If you have not acquired production and safety skills, you need to purchase them. Hiring employees with production expertise and a solid safety foundation is mandatory to staying in a sustainable business with a viable product line or service. Gaps in production expertise invariably lead to client frustration and costly lawsuits. A gap in safety competency could result in someone getting hurt and surely would lead to workers' compensation claims, higher costs of doing business, and possibly to business failure. Someone getting hurt through ignorance is not an excuse in the regulatory or professional world. At the minimum, you must master production and safety issues as a defensive posture for avoidance of future claims.

Business Management

Running a building/remodeling company should generate ongoing sustainable income. Creating processes that generate value for others requires a mastery of certain skills. Only when you have created value are you entitled to some of it. You generally account for this value in the currency of dollars, but you also can accumulate it in pride, self-confidence, and peer respect—to name a few other attributes of success. Each of these nonfinancial rewards provides or supports personal fulfillment, but if profit is absent the company goes under. The specific skills you need to master include planning, human resource management and development, quality control, and goals measurement.

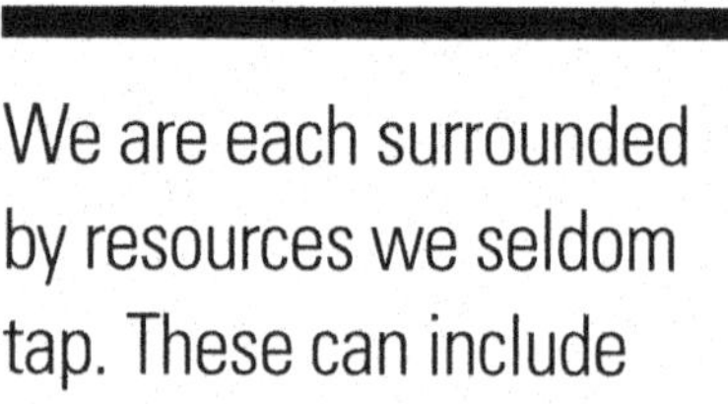

We are each surrounded by resources we seldom tap. These can include our network of acquaintances, financial strength, emotional strength, and personal stamina, to name a few.

Losing money in the production phase of a building project is easy. The production of a home is fun and personally rewarding. However, you need a touch of introspection to ask: "If you are too busy working ***in*** the company to do this critical planning and management, who ***is*** working ***on*** the company?" This situation is functionally equivalent to working for a company that has no boss or leadership. You must master the business management role because it helps to ensure stability and continual income.

Financial Management

Plenty of our small-business peers think financial management involves only keeping cash in the account and balancing the monthly statement. The competency of financial management also includes the skills needed for budgeting, financial analysis, cash flow modeling, and investment-return analysis. You need to plan and track your entire flow of funds into and out of the company, and you need to regularly review that flow for inefficiencies. Mapping the flow of funds can be enlightening as is each of the six competencies. To learn more about the finances of your company, you must be familiar with, if not immersed in, its financial operation.

As you can see from the details of the other five competencies, solid estimating alone does not ensure a successful company. You must address and master each of these other competencies. To enhance your core competencies, you can look to educational resources for skills development. These resources include the items listed in Figure 3.2.

Estimating Barriers

Are the barriers informational or motivational? As you unearth deficiencies in your construction competencies you need to know why you are not fully competent in any given area. As a former teacher and licensed secondary principal by degree, I would regularly (and wrongly) think that builders were simply too deficient in information to become experts. I thought an educator's role was to pass on information, to unlock the doors to a vast repository of unknown facts that would create success. After failing to enhance a few

FIGURE 3.2 Educational Resources for Skills Development

The School of Hard Knocks, Where you never graduate, and you have no curriculum.

Multigenerational Businesses. An existing entity has a strong market advantage and cultivated processes. It offers a solid start if you are lucky enough to be part of such a business.

Personal Mentoring. You can cajole your network of successful peers to provide mentoring or you can purchase it. A simple exercise might be to ask yourself who among your business friends is smarter than you. When you find one with whom you are comfortable, ask this person to look with you at your personal and company strengths and weaknesses and ask if this person is willing to give you a few ideas to work on. I call this person or persons the unpaid corporate board.

Trade Press. Cull the periodicals to which you subscribe for nuggets of estimating wisdom, checklists, tips, pricing trends, anecdotes, or articles of current impact on your estimating skills.

Trade Schools. Generally you can find primer classes in the trades although support for these programs has shrunk in the past decade. Occasionally, these schools provide adult or professional level classes.

Trade Associations. Take advantage of this fine source of personal skills development through peer networking, trade show presentations, seminars, dinner meeting presentations, specialized education, and others.

Internet Sites. The low cost of distribution, variable and updated content, and the global access suggest that this teaching venue will change the industry in the coming decades. The evaporation of industrial arts from public education, in its former role as a standard of content delivery, left a gaping hole in our domestic delivery of construction training. The Web offers a huge potential. To tap into some of its benefits, you should mine supplier and association sites for business and technical content. But respect the copyrights on material you find there.

Supply Chain. The manufacturer, distributor, and retailer support network is alive and well and offers much in the way of material, technical, and installation support. The content derived from the supply chain can provide the basis for estimate numbers, or it can add clear insights into the degree of difficulty in completing installation. The supply chain has always been an estimator's best friend in supplying prices, and it offers a number of other benefits to your company as well. These benefits can include sourcing for trade contractors who can help with current or future jobs that the salesperson may have in his or her client files.

peers' careers while attempting some professional teaching challenges in the trades, I began to realize that mere access and exposure to information was not going to cure those deficiencies. I soon realized that barriers to success can be informational or motivational.

Informational Barriers

These barriers are nearly gone. Dissemination of information is low cost, and information is readily accessible through the supply chain, trade shows, professional journals, the Web, and other sources. Informational barriers are lower than ever.

Motivational Barriers

You can look at the elimination of motivational barriers as an opportunity to increase your estimating skills. Ways to overcome these barriers could include prioritizing the items on your schedule, and allocation of sufficient time to the less-fun activity of preparing estimates.

I know several highly competent builders and remodelers who continue to prepare estimates on a yellow legal pad; long ago, they should have passed the hurdle of computer competency. However, I am self-assured that their method works to meet their needs and that they will not change.

If you are constantly putting estimating work on the bottom of your to-do lists, your customer is likely to move on to someone more responsive. Fear of success or fear of failure may form another motivational barrier. If so, you might do nothing on the estimate because you have work backed up and don't want the additional backlog.

This type of barrier is quite real, and it is not an informational barrier. If you process-mapped the management of information in a paper-formatted estimate, you would instantly see evidence that you could streamline a paper work flow process and that a computer-based system would reduce errors. Using a computer-based system would not diminish the information in the least, and it would eliminate most of the steps in the process of transferring information from one system to another.

Personal Skills: Mastering Estimating

You need a number of personal traits and skills to become proficient at estimating. Some are fundamental, such as good eyesight to be able to see minor deviations in earthen grade or jobsite conditions under close scrutiny. The more nuanced traits are cultivated skills developed over time.

You can hone these practiced skills for quality. Achieving a level of self-discipline to check every line item on a site inspection is not an accident. When the list is long and time short, it is easy to skip some items and move on. It is the omission in estimating that will later haunt profits, so the personal discipline to be thorough can only come from within.

Perhaps as a natural result of years of quality-control work, I find I can barely drive down the street without checking for shutters askew or a leader pipe disconnected that foretells a possible water infiltration problem.

Observation

When doing a site visit for a remodeling job, you walk through a room and only see the space between the walls. A consumer looks at a room with a vision of what is going to occur ***in*** the room. An estimator on a site visit needs to know what materials and labor make up the building systems in the room. The difference would be monumental for the outcome of the estimate. When you are looking at a site, you must turn on these observational skills so you can gather as much information as possible. Do you see evidence of water damage? Do you notice any unusual wear? Are any components out of level?

Observe the general condition of the home upon first entering to determine the owners' possible level of expectations. When you have walked through enough white-carpeted pristine homes, you will understand that the owner of such an abode will likely need the highest standards of cleanliness during the remodeling process, and you will be sure to add some additional dollars to the cleanup line items.

When you are doing insurance estimating, these skilled observations convert to a line item of cost for each occurrence you find. This power of observation is a "Where's Waldo" of the building and remodeling industry for estimators. (*Where's Waldo* is a series of picture books geared to children that invite the reader to look for a small item in a complex, highly detailed picture.) They have a tendency to check technical details when looking at a building because they are often both the estimator on the front end of the job and the quality control agent upon completion. Practice and develop your team members' observational skills simply by asking while sitting on spackle buckets over coffee:

- What is likely to be behind that wall?
- Is that wall straight?
- What is wrong with the electrical work in this room?

Training others to develop a critical eye empowers you, and their observations will broaden your insights as well. Observation is a personal skill that can reap rewards for the proficient and wreak havoc on the unskilled. Cultivate the personal skill of observation as a defensive strategy for estimating success.

When I am adding onto a home, I will often sight the exterior walls, check the reveal around the doors, look for separation of the base moldings, and examine the floor for signs of settlement. These items all indicate possible structural concerns.

Conversation

Many private sector jobs are still awarded with a large degree of personal relationship built into the contract. Laws may limit the public sector to the lowest bidder based on detailed specifications, but private sector consumers assign some degree of credit for

- confidence in the company
- professionalism
- reliance on reputation and referral

To that end, the estimator should engage the client to understand his or her wants and needs beyond the written plans. These wants and needs are not defined or at least not fully defined in the plans; they are omitted in the scope of work; and if they are not accounted for in the line items, they can cost the builder or remodeler a lot of money. In a remodeling job a simple question might be, "Can the tradespeople use the bathroom in the home?" Otherwise you need a line item in the estimate for portable sanitation.

In this early client meeting, you should discuss the availability of financing and the source of funds to pay for the work. If the builder or remodeler should be waiting for a series of large draws, discounted materials and preferential pricing from trade contractors may only be possible with a line of credit, and it carries hard costs. You may need to make a management decision about whether to accelerate construction with a cash infusion that would effectively buy a higher profit on the job. These real financing costs rarely appear on plans or in the specifications.

I often ask seminar attendees if they have a process for the daily mail delivery. Invariably everyone has some process. The fun comes when you track what you actually did with the mail over the past week and document it with a flow chart. Often the actual process involves multiple sorting, second readings, piles, and filing. Once charted, the process invariably improves as you become aware of redundant or wasteful steps.

Accuracy

A sloppy take-off leads to a sloppy estimate. Customers appreciate a neat and thorough estimate that contains an early reflection of your attention to for details. You can demonstrate this accuracy in a thorough investigation of the site, communicate it in a careful take-off, and affirm it in a well-executed job. Being accurate is not an accident. It is a personal trait that takes practice and commitment.

Visualization

Because you build the estimate on paper as you would in the field, one way to look at the site is to visualize it during preparation for construction, in the process of construction, and after the work is completed. A common comment from consumers is that they simply can't visualize a completed project or home from the plans. You should picture the job unwinding before you as if it was in a time-accelerated movie. As you visualize the work in this way, you can see the pitfalls in some of the costs that you need to accommodate in the estimate. If you need to stockpile the excavated fill and prevent soil erosion there, you need to enter a cost beyond merely digging out the soil. If the fill has to go off-site, you will have a trucking cost. If the fill has to be moved to a distant corner of the lot, you will have some machine time involved. Watching the construction work unfold in your head until you see the final home is a skill set that will disclose the pitfalls in the estimate early and allow you to account add line items and dollars to the estimate.

Process Control

Building and remodeling clearly are linear processes with a few deviations to accommodate the preferences of the individual builder's or remodeler's systems. Unless you understand that construction, as well as business, is a work flow, you are resigned to repeat a series of nonlinear tasks. The efficiencies disappear and goals are seldom met with chaotic activity.

In examining nearly any process, you probably will find that you can improve it. A few tasks that can help are to think about what processes you currently have in place for estimating success and to chart the flow of work within the process. How do you get the data to put in an estimate? What would this flow look like? Take the time to put these few steps in a chart, and likely the inefficiencies of your current activity will jump off the page. You may omit such critical steps as interviewing the client if you are bidding architects plans. You may dispense with a site visit if the scope of work is relatively small. These process shortcuts can cost plenty of money and add risks to the creation of profits.

A few years back I discovered drawing software that can help you record and improve your information systems and business processes. Understanding and controlling these processes is particularly relevant to estimating because missing a step can be fatal to profits. Each estimated component builds to the next construction phase. A chart reveals the gaps in the construction process, and these gaps are best filled with dollars in the estimate covering materials and labor costs that may have been overlooked.

Cleanup between trades is an example that comes to mind. After the drywall is hung and before taping you need to have the scrap removed.

Left to their own devices some years back, our spacklers would toss the scrap out the nearest window. Many builders tag the trades to do their own cleanup—a fine solution, but you are likely to have a different set of standards for every person on the job. We keep a cleanup line item in the estimating list and are sure to have the site cleaned between trades and for jobsite safety enhancement as well.

The intent of this chapter on competencies has been to discuss the specific attributes of an effective estimator and to briefly describe how an estimator cultivates these skills. They are well within the grasp of estimators, and each needs to hone them continuously.

CHAPTER 4

Spreadsheet Estimates

A spreadsheet is nothing more than a database of information that can be manipulated with math formulas; it is a place to store information. A spreadsheet is the physical layout of information into columns and rows. This information can be saved, altered, updated, linked to other spreadsheets containing different information, graphed, hidden, templated, or formatted. You can also attach notes to each cell of information, track changes, e-mail, print or fax, insert photos, diagrams, and graphs. You also can sort, filter, export and import data, and introduce all sorts of formulas. Spreadsheets are indeed a powerful tool.

Working in a spreadsheet format for estimating is brilliant and extraordinarily useful; the spreadsheet format powers all estimating software. Creative programmers have built user-friendly checklists to trigger reminders for items that may otherwise be forgotten. They have also built user-friendly screens to input data and templates to retrieve it as memorized reports or presentations. This preprogramming of a spreadsheet into user-friendly input and output forms can be an efficient mechanism for getting your estimates and office functions to run smoothly. Stock estimating software can provide the benefit of the forms design and database creation. The experience of professionals and pioneers can be an invaluable timesaver in developing your own systems. However, commercial estimating software is not a mandatory component of office systems. For example, spreadsheets can readily empower your estimating capabilities. Combined with other software components, your office systems can flow together efficiently.

Within a computer application, you can make multiple uses of the same information you gathered about a construction job. A spreadsheet allows for various ways to adapt this same information and use it to interact with the customer, the labor force, and the supply chain. You do not need to create a new document when each need arises. The basic spreadsheet contains nearly all of the repetitive information, and you can adapt it for these multiple tasks.

An attorney friend once quoted his father who told him to always "Leave a paper trail." A well-documented job file, complete with customer sign-offs will minimize risk. Documentation for change orders, selections, sign-offs, notices, and agreements minimize risk of misinterpretation and allows others to follow your paper trail. The spreadsheet becomes this core of information for the job.

If the only line item currently available to cover risk is the profit line item, you should protect it through documentation that can limit exposure, clarify conflicts, transfer risk to others, and defend profits.

Just as a spreadsheet estimate has multiple uses in presentation, you have as many other reasons to use this electronic format of columns and rows. The most obvious reason is repetition of simple tasks such as retrieval of core information and math accuracy. Filing of the estimates is certainly more convenient as well. You can file the estimates by customer name, job type, or date. Whatever the filing system you use, it should reflect your retrieval needs.

We file contract documentation by current jobs and closed jobs.

You can use jobs with like characteristics to get an estimate started rather than re-create all the items from scratch. Creating and storing some templates for common and similar jobs can save you lots of time and help you to avoid omitting line items. You can save a series of estimates as templates. You might create one each for kitchens, baths, roofing, or siding. If the style of the home to be built or a subdivision of similar homes generates some repetitive consistency you could save a template for these characteristics.

Chronological Estimating

Every phase of construction builds upon the previous work, and no work starts until the local jurisdiction issues permits based on complete drawings. Estimating in a chronological order of construction allows the builder to think through the job fully before putting a shovel in the ground. When thinking through a remodeling job, you discover access issues, explore cutting and patching challenges, and most importantly, you assign costs to these issues. Many software programs group work by trade or product category, not by sequence of work. The rationale for chronological estimating, scheduling, and cash flow information is based on the sequential format of construction.

Once you prepare an estimate, you can use it to draft a draw schedule that replicates a time frame of when the builder or remodeler incurs the costs. The initial deposit should be enough money to get you to the first draw; each draw needs to be enough money to get you to the next draw including overhead and profit. A policy format for stating this situation would be that the amount of deposit should be commensurate with the risk to mobilize and get a project to the next payment. Each ensuing payment should be sufficient to fund work to the next payment, and so on. More about this topic appears in Chapter 10, "Financial Analysis."

The most common mistake in estimating is omission, and you can make this mistake a rare occurrence if your estimate form carries line items for nearly every possible task and some leeway for special tasks. You should organize your estimates in a format that reflects the workflow of the job. Builders seem to be innately capable of doing solid scheduling. Often the first thought on awakening is to listen for raindrops to get a perspective on the weather for the day. Thoughts run to who has to be where first thing in the morning, then a look into activities to occur later in the day, followed by a glimpse of the evening calendar. These animated steps lead to creating a schedule. These practiced techniques of managing a day can carry forward to managing an estimate, creating profitable work flow, and ultimately building a profitable business. Because contractors are planning their days and work chronologically, they should organize their estimates chronologically. The orderly work flow of a building project is a logical format for organizing an estimate.

Building an estimate is like building the job on paper. Building the job in your mind can help you find construction problem spots, access issues, technical conditions, and a host of other possible cost items you need to account for in the estimate. This dry run at construction on paper is invaluable. A solid foundation of an estimate is chronologically ordering the sequence of work. What needs to be done first? What is done next? The animation of this construction work through chronological thinking and chronological estimating helps to ensure that you do not overlook any steps. Many builders and remodelers agree that the single most frequent error in estimating is not getting a price wrong but omission of a task or line item.

The key is to break the complex process of a building project into smaller pieces or tasks. Estimating breaks the job into clear line items.

For you to build this chronological thinking into a checklist of tasks is imperative. You define each task by its scope of work or size and assign a price within the spreadsheet. More information on getting these numbers will follow in Chapter 5, "Seven Ways to Get the Numbers." Given the prevalence of spreadsheets for math calculation, putting this list of tasks into a spreadsheet that completes all the math is simply good business. Some contractors still rely on a yellow pad, pencil, and a hand calculator. Those days should be gone along with the #4 finish nail and nail set punch for doing wood trim installation.

Start with the first steps in the contracting sequence. Generally these steps involve the drawing and engineering work followed by permits. You may also need legal and appraisal work, and you will have to accommodate hard costs early in the sales process. These costs could include design costs and duplication of plans or renderings. Each of these tasks becomes a line item. These defined tasks should follow in the same column and populate a series of rows. The columns to the right of the items can be used for pricing, quotes received, notes, details, pricing variations, or organizing subsets of tasks. All the tasks contained in site work, for example, can be subtotaled to better understand the overall lot improvement costs. You can also use these subtotals more readily later to compare your estimated costs to actual costs. A short, generic spreadsheet for an estimate appears in Figure 4.1. The purpose is to show the chronology of work flow as well as the introduction of multiple columns to add information. In this example, you can calculate the work as materials and labor or as a single subcontracted price.

You may need to define some tasks early and well to clarify the scope of work for sales purposes and later to document a more complete understanding with the customer. The cells containing notes can be more important than the prices that you insert because they define scope for that item. Suppose you defined the line item of closet interiors to be "as selected by the

FIGURE 4.1 Short Form Spreadsheet

Short Form Estimate

JOB NAME	JOB NUMBER	DATE
LOCATION	JOB MANAGER	CHECKED BY
ESTIMATOR		ENTRY BY
JOB DESCRIPTION	*DENOTES FIRM BID	DATE

Phase	Description	Vendor	Labor	Material	Other	Subcontract	Total
1	Plans & permits						0
2	Demolition						0
3	Site work						0
4	Excavation						0
5	Concrete						0
6	Masonry						0
7	Framing						0
8	Roofing						0
9	Ext trim						0
10	Windows/doors						0
11	Siding & gutters						0
12	Plumbing						0
13	HVAC						0
14	Electric						0
15	Insulation						0
16	Drywall						0
17	Ceilings (Drop, etc.)						0
18	Millwork/trim						0
19	Cabinets/tops						0
20	Floors						0
21	Painting						0
22	Debris/cleanup						0
23	Landsc./paving						0
24	Misc.						0
25	Supervision						0
26	Decks/patios						0
27	Allowances						0
28							0
	Subtotal	0.00	0.00	0.00	0.00	0.00	0.00
	Overhead	0.00	0.00	0.00	0.00	0.00	0.00
	Profit	0.00	0.00	0.00	0.00	0.00	0.00
	Total Costs	0.00	0.00	0.00	0.00	0.00	0.00

TIME

owners" and were to insert a price of $500. The price would likely be plenty of money for wire shelf and pole combinations. It would not be enough money if the customer wanted hardwood shelves, storage bins, rollouts, and by the way, a jewelry drawer. You just lost $2,000 at the line item by fulfilling your own specification that the closet interiors had to be built "as selected by owners." Can you blame them at installation for asking for what they want? The answer is no, and the blame goes to the contractor for having a process so deficient that nobody knew what was going into the closets until the carpenter went to measure them.

The mere existence of the line item should prompt some discussion about the scope of work far earlier in the process. To defend against this $2,000 loss at one line item, the wording in the notes cell could read: "wire shelf and pole combination at 6 feet a.f.f." This scoping of the work defines your cost limits. This cost item is now only slightly variable depending upon who is doing the installation. A wire shelf-and-pole combination has quite a narrow range of materials cost, and you have set the installation at a configuration of one wire shelf set at 6 feet above the finished floor (a.f.f.). No one should question what is expected, and your defensive notes just minimized the risk of this line item's impact on profitability.

To restate this concept, you control the risk by narrowly defining the work rather than by padding the number at the line item. Sometimes people in the building and remodeling industries look at a complex job and "throw" a high number at it in an attempt to bring parity with scope and price. This irresponsible estimating is not likely to get you the work, and it will certainly leave all the risk in play because you have not defined the scope of work even though you have escalated the price. You need to make a prudent effort to break the complex project into components and get real numbers based upon a clear scope of work. Avoid using the defensive tactic of adding cost to avoid the work of getting clarity on the scope of work.

As the information changes and becomes clearer, you need to appropriately update the estimate. These changes could be details added by the customer, specifications added by the architect, code compliance issues from the building department, or new information gleaned from additional input no matter what the source. An estimate is always a work in progress that reflects your understanding until you commit to the defined scope of work for a fixed price by entering a contract. Once the home buyer and you sign the contract, the base contract is fixed. Thereafter, any deviations from the scope ***require*** a written change order.

The Steps to Build the Job

When starting the actual construction work, you almost always find some tasks for the job-mobilization tasks, such as site protection for a new home, floor protection in remodeling, or simply moving equipment. Each task needs to carry a price into the estimate. The subsequent items in the spreadsheet should reflect the sequence of work at the jobsite.

The many steps to the reconstruction process include cleaning, moving, and storage of contents, as well as all of the subsequent construction activities. Characteristically, the insurance estimating process requires a high level of detailing. From a builder's, remodeler's, or trade contractor's perspective, if the unit price on labor or materials is too tight, and the insurance industry standard limits of 10% and 10% (of overhead and profit allocation) are in play, the way to add some defensive cushion to the building contract is to further expand and define the priced line items. This expansion and definition could involve enhanced mobilization, or you could expand the steps in the process and add the cost to each expanded activity.

A comparative example might be that in a building contract you may have a line item for interior painting accompanied by a price. The same estimator for insurance jobs may estimate line items for site protection, drywall preparation, priming, materials assembly, materials storage, temporary heat, finish coats, and final cleaning and removal of debris from the site. Both estimates get the room painted, and one lists more details in the break out of activities.

The enhanced line items at least give the builder or remodeler a discussion platform when the time comes to meet with an insurance adjuster and defend your estimates against the carrier's expectations.

Multiple Uses for a Templated Spreadsheet Estimate

You can use the same template for an estimate for multiple purposes. The core information remains the same and does not need to be reentered. Every repetitive step you eliminate is a cost not expended. The client becomes familiar with these repetitive uses, so you minimize uncertainty and confusion. Several of the uses tie directly to the estimator's data gathering. Repurposing the same checklist for data collection ensures that the site visit is productive. Using the same list for a client interview suggests that you will glean information on each topic if the owner has special needs.

Client Interview

You can use a blank estimate template early with potential clients who have been prequalified to justify additional time in the sales process. Rather than engaging the client in an informal exchange of ideas, you can walk the client through the chronological steps in the process and harvest their input to be sure that your estimate and the work meet with the client's expectations. This layering of input into one core document will solidify some of the details to be entered into the spreadsheet. Any number of additional details may come up in the interview process that will require you to add an activity line item and assign a cost to the work.

By asking your clients to add their thoughts to each line item, you can expedite and focus the on-site interview. The alternative is often a rambling unfocused social visit with a slight overlay of contracting process rather than this crisp step toward accuracy in the estimate. You can modify this form through straight deletions, hiding columns or rows, and only delivering a fax or hard copy that the customer cannot manipulate.

Data Collection

You can carry a printed copy of the form to any of the project meetings and use it to gather pertinent data at every stop. The municipality may have some cost-related requirements from any number of departments that regulate construction. The local municipal engineers recently were empowered to require storm-water detention facilities for existing homes when an addition is built. The intent is to recharge the ground water. Water driven down by gravity encounters the lawns of existing homes. The lawns slow sheet runoff and generally capture rainwater anyway, but that fact matters little to the municipal engineers. This encumbrance must be designed by an engineer ($1,500–$5,000), permitted ($100–$500), installed ($2,500–$5,000) and connected to all existing leader drains ($2,000–$4,000), inspected, then the entire disturbance needs to be regraded and seeded.

I often send a slightly modified version of the estimate template to the client early to get the client's feedback into discussion. Before I release the form to the client, I take out the costs that are not otherwise a part of the discussion, as well as remove any references to soft costs, overhead, warranty costs, or profits because they might trigger issues that, at this point, are secondary to the scope and price of the work.

At one time, a building permit was handled out of the building department. Though this procedure makes plenty of sense, today the building department is only one of a large number of steps to get through from zoning, engineering, the municipal road department, health department, water department, ad nauseam. The state of New Jersey has 146 different permitting bodies regulating construction. Many do not often apply to a specific job, but you surely must address the costs of compliance lest you omit something that you must pay for later out of your profit line item. Regulatory risk is abundant.

Site Visit

During a site inspection, this spreadsheet becomes a checklist for items to be reviewed. Each line item triggers a look to see if that item involves any complicating factors. When a basement is to be finished and a bathroom installed, the sewer-connection line item should remind you to see if a floor-level drain exists or if a drain goes out through the wall. Either way you will need to cut some concrete. However, if the waste line goes through the basement wall, you will need to include a pump system in the cost estimate. Every home has a sewer or septic waste line, but knowing the height of the discharge will drive the scope of work and, therefore, the cost.

When you are looking at a roofline, the depth of shingle along the rake edge can tell you how many layers are on the roof. During the 1950s and

1960s, less so in the 70s, adding asphalt shingles over a wood shake roof was standard practice. The codes accommodated this practice by allowing the shakes as sheathing. Now, if you have to strip the roof and are looking at asphalt shingles on top, cedar shakes might be underneath them. You will need to remove them and sheath the entire roof in plywood.

Determining the presence of sheathing may not be a conventional inspection, but omitting this step may have you buying a lot of plywood without being compensated. In addition to solid inspection work, several clauses can help mitigate risk. You may want to add appropriate detail, such as, "All reroofing quotes anticipate functional sheathing on the home. Rot, replacement, or an insufficient surface shall trigger a time-and-materials change order for repairs."

Often I will ask a lead carpenter to participate in a site visit when the contract is imminent so more accuracy can go into the contract reflecting the actual work.

Staff Estimate Discussion

Regular feedback from the people doing the work will help you to isolate costs. They may see what the estimator misses. During a look at one whole house remodel where the floors were to be raised, nobody else caught that the firebox would no longer work and would need to be rebuilt. This work was neither in the plans nor on the specifications. When you change a roofline, often you need to raise the chimney. The architect's catchall phrase of "match existing" often means more to the field crew than a simple square-footage takeoff on the estimator's desk. More input generally increases overhead slightly, but it is well worth the investment in accuracy and involvement of the team that has to build the work. You can use the same list as a trigger for discussion to solicit input from trade contractors.

Proposal

You can repurpose the same estimate form for delivery to the customer with a few keystrokes. You need to enter the details as notes and list the items in the left column. You include your totals for cost at the bottom of the estimate. Spreadsheet software allows for hiding columns or rows, so you can hide the numbers that relate to estimated costs. The information is not lost; it merely disappears from sight.

Even though you collapse the numbers from view, the items, notes, and details remain visible, and this document can now serve as a written proposal for a defined scope of work. You can replace the word *Estimate* in the title with the word *Proposal,* and it is ready for delivery and acceptance. This continued use of the same document eliminates the possibility of the estimate and the proposal conflicting. They are one and the same. If the numbers appear in a column, you can hide them.

You may want to hide any reference to overhead, profit, or callback costs that you may build into the estimate. You also can hide titles in rows. These items can raise questions that may not have a palatable answer. Many prospective home buyers think builders and remodelers should work for nothing. Showing such customers that profit is in the budget may be a red flag.

Another sound security tip is to send your estimates via some delivery method other than e-mail, unless you eliminate the sensitive information and re-save the estimate as a separate presentation document. You would only have to make this mistake once to have your costs, mark-up, and overhead contributions disclosed.

Some customers respect our need to generate stability through profits and would not touch the subject if it appeared, but I can think of no good reason to lead with these line items exposed and suggest you hide them. An example appears in Figure 4.2. I have formatted the columns and rows with a background screen so that you have no chance to send out an estimate with prices disclosed or sensitive information revealed. We use yellow to highlight the cells that we do not publish to the client. It makes the lines that need to be hidden before presentation easy to see.

Payment Schedule

Yes, you are still using the same spreadsheet to get a fair and equitable payment schedule developed. Because the line items outline the work chronologically, the costs also fall into line chronologically. So if you draw a line in the estimate up to the completion of the foundation and add the costs above that line, the amount of money to get to that progress point is a hard number. You should declare this number plus your percentage for overhead and profit as the deposit required to begin the job. You can state in the draw schedule that you deem the costs incurred up to the next progress payment (roof framed) due for payment at the start of roof framing.

You can deem the costs incurred to get the mechanicals installed due to be paid at the start of that task. The effect of this clean analysis is that you are working with the client's money. The old-line adage of 10% down, 50% when half done, and the balance (40%) on completion has no rational basis. Nor do three payments of one-third each have any logic. These practices simply have no tie-in to the costs incurred.

If you place activities in the spreadsheet for ordering cabinetry or special orders, these items incur cost liability, and you should collect from the customer for them. This system allows for as many payments as the job may warrant based upon job size (Figure 4.3). It keeps the builder's or remodeler's money out of the job. If you experience a need for accelerated cash flow for deposits and orders, you can insert these activities as items earlier in the estimate. An example of this idea appears later in this chapter, the estimate separates rough and final plumbing. You can spread these items throughout the layout so their work position is reflected in the spreadsheet. Conversely, if left as shown, the completion of rough plumbing and the beginning of

I am not an advocate of using supply channel funds, trade contractor good will, or extended terms to get a job built. The customer should pay for the job.

FIGURE 4.2 Converting the Estimate to a Proposal

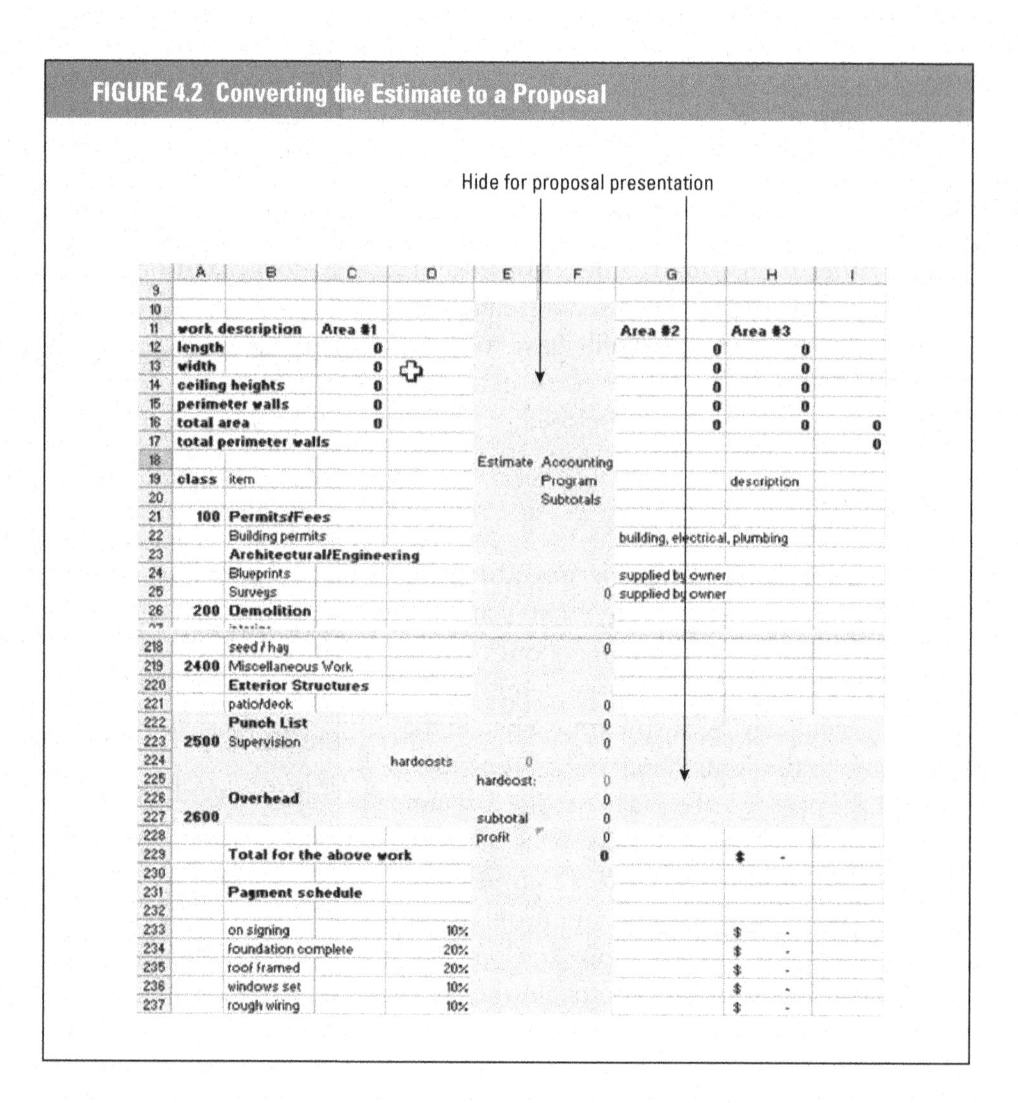

	A	B	C	D	E	F	G	H	I
9									
10									
11	**work description**		**Area #1**				**Area #2**	**Area #3**	
12	**length**		**0**				**0**	**0**	
13	**width**		**0**				**0**	**0**	
14	**ceiling heights**		**0**				**0**	**0**	
15	**perimeter walls**		**0**				**0**	**0**	
16	**total area**		**0**				**0**	**0**	**0**
17	**total perimeter walls**								**0**
18					Estimate	Accounting			
19	**class**	item				Program		description	
20						Subtotals			
21	**100**	**Permits/Fees**							
22		Building permits					building, electrical, plumbing		
23		**Architectural/Engineering**							
24		Blueprints					supplied by owner		
25		Surveys				0	supplied by owner		
26	**200**	**Demolition**							
218		seed / hay				0			
219	**2400**	Miscellaneous Work							
220		**Exterior Structures**							
221		patio/deck				0			
222		**Punch List**				0			
223	**2500**	Supervision				0			
224				hardcosts	0				
225					hardcost:	0			
226		**Overhead**				0			
227	**2600**				subtotal	0			
228					profit	0			
229		**Total for the above work**				**0**		**$ -**	
230									
231		**Payment schedule**							
232									
233		on signing		10%				$ -	
234		foundation complete		20%				$ -	
235		roof framed		20%				$ -	
236		windows set		10%				$ -	
237		rough wiring		10%				$ -	

rough heating, ventilation, and air-conditioning (HVAC) will trigger a final plumbing cost to be paid as well.

Statement

A statement can be built directly from the spreadsheet by simply mimicking the draw schedule in Figure 4.3. You can hide information other than the letterhead, customer information, and the draw. The reason to hide rather than delete the numbers is that the formulas will stay intact. Alternatively, you need to cut the data and paste it back as a numerical value rather than a formula. The formula references data that you would have just deleted. You can add rows to credit the customer when he or she makes payments. You can add rows for change orders and allowance reconciliations (Figure 4.4).

If this statement is part of a larger spreadsheet file, you still can't e-mail it to the client because the other information would be vulnerable. If you save the statement as a separate file, you can then e-mail it for collection. The linkage to the contract and estimate keeps a flow of styles, accuracy of data, and consistency of file storage.

FIGURE 4.3 Converting the Spreadsheet to a Draw Schedule

Draw Schedule = Estimated Cash Flow Needs **Short Form Estimate**

JOB NAME	JOB NUMBER	DATE
LOCATION	JOB MANAGER	CHECKED BY
ESTIMATOR		ENTRY BY
JOB DESCRIPTION	*DENOTES FIRM BID	DATE

Phase	Description	Vendor	Labor	Material	Other	Subcontract	Total
1	Plans & permits						0
2	Demolition		DEPOSIT				0
3	Site work						0
4	Excavation						0
5	Concrete		Draw #1				0
6	Masonry						0
7	Framing						0
8	Roofing		Draw #2				0
9	Ext trim						0
10	Windows/doors						0
11	Siding & gutters		Draw #3				0
12	Plumbing						0
13	HVAC						0
14	Electric						0
15	Insulation		Draw #4				0
16	Drywall						0
17	Ceilings (Drop, etc)						0
18	Millwork/trim		Draw # 5				0
19	Cabinets/tops						0
20	Floors						0
21	Painting						0
22	Debris/cleanup		Final Payment				0
23	Landsc./paving						0
24	Misc.						0
25	Supervision						0
26	Decks/patios						0
27	Allowances						0
28							0
	Subtotal	0.00	0.00	0.00	0.00	0.00	0.00
	Overhead	0.00	0.00	0.00	0.00	0.00	0.00
	Profit	0.00	0.00	0.00	0.00	0.00	0.00
	Total Costs	0.00	0.00	0.00	0.00	0.00	0.00

FIGURE 4.4 Converting the Estimate into a Statement

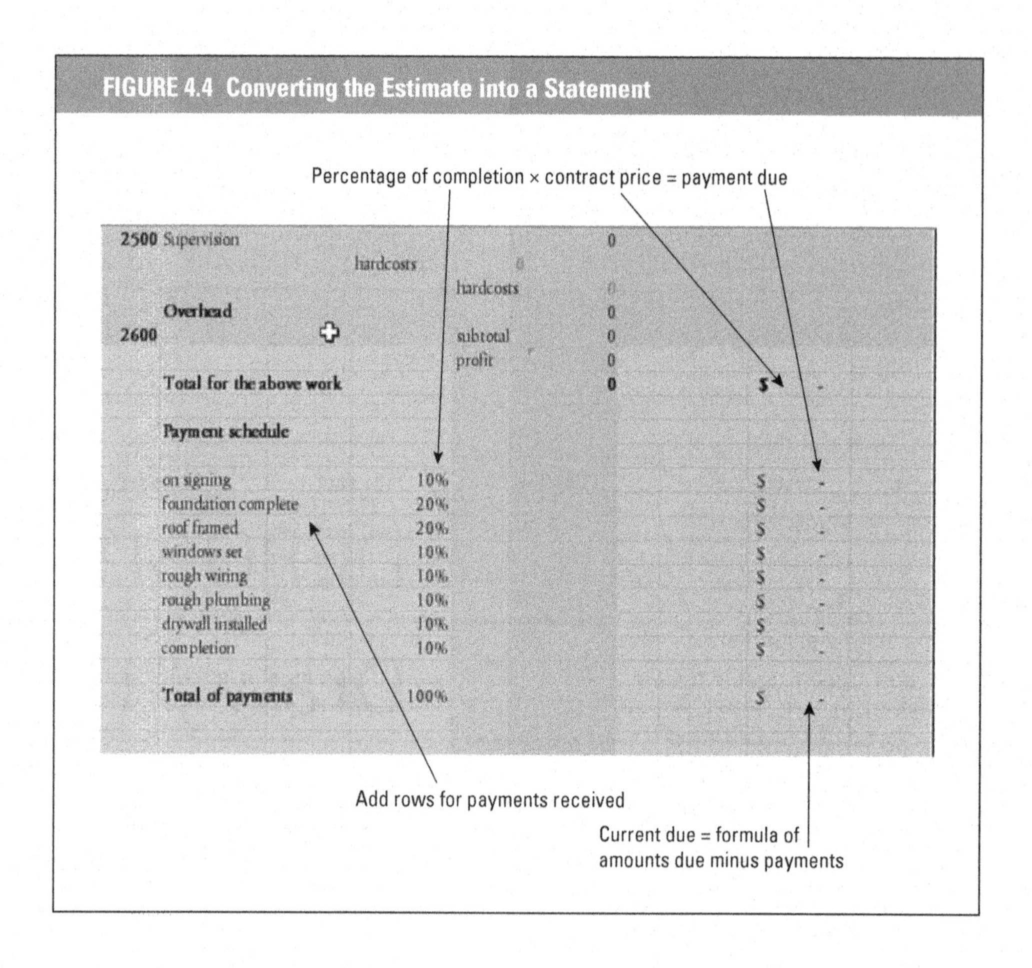

Change Orders

You should view a change order as another small job. You need to estimate change orders as you would do any other job, and it should contribute to overhead and profit production. Using the same estimating form allows all the core data to stay in place. The risk of a change order creating contract chaos is no different from such risk with the core contract. You need to adequately define, price, and present the change order. After the owner accepts and signs it, you complete and bill the work.

You should add an item to the template for change orders to fill in how many days this change order extends the completion date for the base contract. Many builders and remodelers forget this simple step and commit to more work without incorporating a contract extension. When the job starts to drag because of multiple change orders, you need to have the clients' signatures in hand to prove that they delayed the original completion and that they accepted such responsibility by signing the change order with an extension spelled out.

Chronological Job Analysis: Scheduling

A spreadsheet built in chronological order can add not only numbers but also days (Figure 4.5). If you are building a schedule for a job, you can project the

(Continued on page 47)

FIGURE 4.5 A Full Spreadsheet

			Company Name				
			Company Logo				
			Company Name			Date	
Estimate						Company contact	
	Client					Phone	
	Address					Plans	
	City					Architect	
	Phone					Revisions	
	Cell phone						
	E-mail						
Work Description		**Area #1**				**Area #2**	**Area #3**
Length		**0**				**0**	**0**
Width		**0**				**0**	**0**
Ceiling Heights		**0**				**0**	**0**
Perimeter Walls		**0**				**0**	**0**
Total Area		**0**				**0**	**0**
Total Perimeter Walls							
				Estimate	Accounting		
Class	**Item**				**Program**	**Description**	
					Subtotals		
100	**Permits/Fees**						
	building permits					building, electrical, plumbing	
	Architectural/Engineering						
	blueprints					supplied by owner	
	surveys				0	supplied by owner	
200	**Demolition**						
	interior						
	exterior						
	plumbing						
	electrical						
	structural				0	brace exterior walls	
300	**Site Work**						
	Utility Connections						
	temporary electric						
	gas service						
	electric service						
	telephone service					by N.J. Bell	
	other utility connection						

(Continued)

FIGURE 4.5 A Full Spreadsheet *(Continued)*

	individual wells					remains
	water service					remains
	septic system					remains
	sewer service				0	
400	**Excavation**					to 42" for footings
	backfill					rough grade and swale
	move fill					
	trucking				**0**	
	footing drains					
500	**Concrete**					#3000 mix footings
	formwork			0		trench footings
	slabs			0		4" thick - #3000 mix
	structural frame					
	stairs					
	precast decks and walls					
	garage/carport slab					
	concrete labor				0	
600	**Footings/ Foundations**					trench footings
	cement			0		#3000 mix
	concrete blocks			0		
	gravel					4" clean stone below slabs
	sand					
	rebar and reinforcing steel			0		
	other foundation materials					4 mil poly vapor barrier
	labor footings/foundations			0		
	radon					venting to code
	Masonry					
	Masonry Materials					
	chimney/fireplace					
	brick veneer					
	brick/stone wall					
	masonry flooring					
	Waterproofing			0		asphalt coating below grade

FIGURE 4.5 A Full Spreadsheet *(Continued)*

	Termite Protection			0	0	fiberglass sill seal
	Structural Steel					
	stairs					
	beams					
	posts					
700	**Framing**			0		
	Framing Material			0		all framing materials: Douglas Fir
	Floor Framing					crawl space
	posts					
	joists					2x10
	plywood					5/8 CDX
	ledger strips					
	hardware					
	cross bracing					bridging
	Partition and Wall Framing					
	studs			0		2x4 wood studs
	plates			0		treated lumber at masonry
	bracing			0		
	sheathing			0		5/8" at roof & 1/2" at siding
	hardware					
	Roof Framing					
	trusses					
	Gables					2x8 rafters
	ceiling joists					2x6 ceiling joists
	sheathing					5/8 plywood
	bracing					
	stairs					
	hardware					
	Basement Framing					
	stairs					
	studs					
	plates					treated plates
	bracing					cross bridging @ 8' or midspan
	hardware					
	Framing Labor					
	floor			0		
	partition & wall			0		
	roof					
	basement				0	

(Continued)

FIGURE 4.5 A Full Spreadsheet *(Continued)*

800	**Roofing**					
	roofing materials			0		
	metal; roof edgings and flashings			0		aluminum at all roof intersections
	roofing labor			0	0	
900	**Exterior Trim**					
	cornices/rake trim			0		
	soffits: gables flashings					
	misc. trim				0	frieze, rake, corner boards, gable vents
1000	**Windows & Doors**					
	windows					
	skylights					
	storm windows/doors					
	exterior doors					
	interior/closet doors			0		
	sliding glass/French doors			0		
	garage doors					
	hardware					
	installation			0	0	
1100	**Siding**					
	posts and columns					
	siding			0		match existing
	shutters					
	gutters and downspouts			0	0	seamless white
1200	**Rough Plumbing**			0		to code
	Finish Plumbing					copper feeds, PVC drains
	tub					
	shower pan					
	toilet/bidet					
	sinks			0		
	dishwasher					
	water heater					
	laundry tub					
	fittings					
	labor			0	0	
1300	**Rough HVAC**			0		
	Finish HVAC			0		
	furnace					
	thermostats					
	air conditioner					

FIGURE 4.5 A Full Spreadsheet *(Continued)*

	duct work					
	hardware					
	labor			0	0	
1400	**Rough Electrical**			0		to code
	Finish Electrical					
	fixtures					
	labor				0	installed per code
1500	**Insulation**					
	foundation/basement			0		
	roof and ceiling			0		R-30
	wall			0		R-13
	floor					R-19
	weatherstripping and vapor barrier					
	labor			0	0	
1600	**Drywall**					
	material			0		1/2" drywall
	labor			0	0	tape, spackle, & sand smooth
1700	**Ceilings and Coverings**					
	grid system					
	ceiling tiles					
	wall coverings					
	Ceramic Tile					
	tile			0		
	installation					
	shower doors				0	
1800	**Interior Trim**					paint grade
	moldings					
		base				
		chair				
		ceiling				
		installation				
	paneling					
		installation				
	closet					
		shelving				
		hardware				
		installation			0	
1900	**Cabinets and Vanities**			0		
	kitchen cabinets					
	countertops					

(Continued)

FIGURE 4.5 A Full Spreadsheet *(Continued)*

	vanities					
	hardware					
	installation			0		standard installation included
	Appliances					
	range					
	range hood					
	disposal					
	dishwasher					
	refrigerator					
	washer/dryer					
	microwave				0	
2000	**Flooring**					
	resilient flooring					
	carpeting					
	hardwood			0		
	installation				0	
2100	**Exterior Painting**					by owners
	paint/stain					
	labor					
	Interior Decoration					by owners
	paint					
	wall covering					
	installation				0	
2200	**Cleanup**					
	debris removal					
	dumpsters				0	
2300	**Landscaping**					
	paving					
	seed/hay				0	
2400	**Miscellaneous Work**					
	Exterior Structures					
	patio/deck				0	
	Service Work				0	
2500	Supervision				0	
				hard costs	0	
				hard costs	0	
	Overhead				0	
2600				subtotal	0	
				profit	0	
	Total for the above-listed work				**0**	**$**

FIGURE 4.5 A Full Spreadsheet *(Continued)*

	Payment schedule				
	on signing	10%			$
	foundation complete	20%			$
	roof framed	20%			$
	windows set	10%			$
	rough wiring	10%			$
	rough plumbing	10%			$
	drywall installed	10%			$
	completion	10%			$
	Total of Payments	100%			$
	Start Date				
	within 10 days of issuance of municipal permits				
	Completion Date				
	14 weeks thereafter				
	This proposal is valid for 15 days				
	for <Customer's Name>				
	for <Your Company Name>				

duration of each activity, and by adding the number of workdays to the prior start and completion dates, you can build a simple schedule. A cut-and-paste of the activities into scheduling software is the easy solution to detailing a schedule, but you certainly can be also do it within the core spreadsheet document. You can use this same schedule to track delays, order long-lead items, verify work is current, and create work orders that match jobsite needs. Estimating duration is often more difficult than estimating cost, materials, or risk. Some states require an estimate of duration as a contract item confirming a start and completion date.

CHAPTER 5

Seven Ways to Get the Numbers

This chapter discusses the various methods of assembling the hard numbers. Each line item in the estimate has a preferred way to create the appropriate number. The risk of error changes with how you fill in the costs for each item. Use the wrong method, and the risk of using a distorted price goes up. Every cell does not have a single method for getting the right number in it.

You have seven ways to obtain the numbers that fill out the estimate spreadsheet:

- Lump-sum pricing
- Allowances
- Linear footage
- Square footage
- Contract prices
- Day rate
- Quantity takeoff

Discussion of the Methods

A number of these methods correspond directly to the size of the project or its specific dimensions. If you enter simple multiplication formulas into the spreadsheet, you can automate these relationships of size to price to complete a considerable amount of the estimate in a relatively stress-free manner. You draft the embedded formulas so that they automate the process by multiplying the unit sizes times the going street price for the item. A commercial estimating software package hides these formulas from the user. If you need to alter the formulas, the commercial package will provide a geographic regional multiplier, a degree-of-difficulty multiplier, or simply leave a space where you can enter an up charge.

Spending time here discussing how to build these formulas and how to link workbooks within a spreadsheet may be prudent. However, given the divergence of possible spreadsheet software, you can use your software's help menus and instructions. The working application of using certain keystrokes to multiply is, therefore, intentionally omitted for brevity and relevance. For Microsoft Excel users, Jay Christofferson's book, *Estimating with Microsoft*

Excel, 2nd edition, provides that information,[2] and his estimating software, *EstimatorPro*™, supports the spreadsheet approach.[3] Once you understand the concept, and the process of moving collected data into an estimated price, you will find that understanding is far more valuable than to reproduce the actual configuration of the cells and embedded math in this book.

When you are starting to fill in one of the estimate templates, you can begin by verifying the size of the rooms, the perimeter walls, and the ceiling heights. These few dimensions can go a long way toward filling in the numbers. The perimeter wall lengths will yield prices for footings, concrete block, siding, exterior insulation, and sheathing. The spreadsheet can link the sizes of the rooms to the amount of material needed or the labor pricing. This information is hidden within the cells or kept separately in the spreadsheet, and simple math will give you a total. These prices can be left hidden in the cell or obscured on a separate worksheet with a priced database of costs.

You can imbed a number of industry shortcuts in the pricing cell that will get you to a number quickly. For example, 3.7 times the area of a room will generally yield the number of square feet of wallboard surface, which includes the ceiling area. You can then use this square-foot number in the drywall line item to multiply by a cost per square foot. The same area of wall number can be used in the painting line item: again by multiplying by a price per square foot. Some builders and remodelers use 4.0 as the multiplier of the square footage because it produces a figure that is about 8% higher.

This particular formula anticipates an 8-foot ceiling height so you would need to adjust the multiplier to account for deviations from that height. A 10-foot ceiling is 25% greater at the wall area, but the ceiling footage is actually the same. Here one entry for square footage fills in two line items. This effort produces a good return on your investment in data collection time. For example, if the job has 1,500 square feet of floor space and the going rate for drywall is $1.50 per square foot supplied, installed, spackled, and ready for painting, what would the embedded price look like? We know the ratio of wall surface to floor is 3.7 square feet of wall and ceiling per square foot of floor and a street price of drywall in a residence may be $1.50 square feet, so the formula would read (1,500 × 3.7) × $1.50 = $8,325.

You can leave this formula in the cell, and the formula can pick numbers out of other cells. Once you establish the area of the space, the spreadsheet will populate the drywall number automatically. If you put the square footage of the room in cell address B8, the formula would read: (B8 × 3.7) × $1.50. This autocalculation will stay in the drywall estimate cell. You can reuse the formula in the painting estimate cells with a unit-cost adjustment for painting closer to $0.50, and it would read (B8 × 3.7) × $0.50 per foot.

Start by working on the specifications in an estimate that will give the estimator a better feel for the overall scope of work and standards of workmanship expected. An example of specifications driving other costs is as follows. If the trim is to be oak and stained, you need to increase the labor cost for trim installation because the trim carpenter's extra time for the tightest joinery and no exposed glue, and certainly no caulk, raises the cost of the trim installation.

You could make a similar point about painting. If the walls are to be an eggshell finish, you may need to account for additional care with spackle, and extra wall prep may be in order, thus raising some costs in the estimate.

Scope of work can be widely divergent for every line item. One estimator might read roofing work as a per-square straight installation, whereas the customer is looking for the house to be draped with tarps, the plants to be shielded with plywood, and the job to be walked daily with a magnet to remove nails from the lawn areas. Both are accurate, but the two perceptions are on their way to a collision. Each interpretation has a different set of costs. Early definition of scope helps the customer understand what he or she is getting and you to know to know what costs to include in the estimate.

Despite the best efforts of the designers, the architects, and the structural engineers, inevitably some different interpretations of plans still occur. The same could be true for a specification book for the work. Although meant to add clarity, some interpretation may be inevitable. You might suggest filling in the notes column first because the notes help define scope of work. By providing notes, you give the estimator some control over the scope of work and over price. Controlling the scope of work is a critical step in defensive estimating.

I suggest to customers that we can complete nearly any job for any price, if I control the scope. Do you want a family room addition for $1,000? We can do it. I control the scope so the room is a nylon rope (structure) with a blue tarp (walls and roof) over it. The space sheds water, provides for habitation, and has natural lighting features that others may spend thousands to achieve. Quickly everyone understands that scope of work defines the end product and drives all prices, depending upon materials and construction details.

You can use a column to the right of pricing for defining the scope and adding details that communicate to the customer what you anticipate building. These notes and specifications can be as detailed as necessary to be clear and specific. Any vagueness in the scope of work is not likely to consistently fall to the least cost interpretation. Customer's expectations are unbridled by a lack of product definition or limits, and you have no way to control what they expect unless it is defined.

You need some contract language to keep the specifications written by the estimator as the highest in priority in the event of a conflict in the documents. You will find further discussion of this sample language, called the *precedence clause,* in Chapter 11, "Defending the Profit Line." The America Insitute of Architects contract documents may defer to the architect's plans being superior to any other documents. In the plans, you should have a statement that hard dimension numbers take precedence over scaling the drawings, and that dimensioned floor plans take precedence over elevations. A customer having told you once about a certain item should not have precedence over the written proposal. You need to do this scoping of the work first (before pricing begins) lest you end up with a long list of uncertain items as the foundation for a contract.

The excerpt from a spreadsheet in Figure 5.1 has two columns screened. This screening provides a reminder that you can later hide these cells so you can collapse the form and present it to the customer. In this way, you avoid double entry—from the estimate to a presentation format—that increases the chance of errors. You can tally the costs internal to this spreadsheet for each section of work. This possibility is helpful for job costing and for projecting cash flow, and they in turn help to create balance for a draw schedule.

An acquaintance some years back defined the scope of the tile line item with the following words: "As selected by owners." Oops . . . need I say more? The customer first selected tile that had to be set in wet cement, a classic "mud job" that is very time consuming. They then added a pattern and custom tiles, in an unexpected wainscoted wall no less. By the time a seat, soap dish, accessory shelf, and stone threshold were added, the poor builder was talking to himself. This situation occurred because he failed to defend profits with a well-defined scope of work at the tile line item.

Lump-Sum Pricing

Lump sum (meaning grouped together) is a simple way for you to get some numbers into an estimate without having to break apart the entire line item to get to the smallest single components. The estimator should decide which lines he or she can aggregate without risk.

Can you add $300 for nails in stick framing a house, or do you need to know how many nails of which size the house will require? If you are a full-time framing contractor, you should know how many of each type you would need and how many you have in inventory. This knowledge may split how many you would order from how many you will need at the jobsite. Whatever the source of materials, you need to include the cost of inventory in the estimate.

A lump-sum price is likely to be sufficient for cleaning the job on completion. If you include some reasonable experience-based number for final cleaning, the level of breakdown for the cleaning supplies will neither make nor break the success of the job. The use of lump-sum pricing forestalls the need for finely detailed materials lists until you order the items. The estimate line item is not meant to be a work order to labor nor a materials list for ordering, but a representation of the cost to keep budgets accurate. A lump-sum price for carpentry trim labor might suffice rather than a full work list of every piece of molding by footage and an installation time per piece calculated.

Allowances

An allowance is a fixed or a unit price included in the estimate that limits the risk to the contract profit from selections that exceed the estimators' expectations. An allowance is a wonderful estimator's crutch. It protects the estimator against price overruns while it accommodates the client's lack of selections for some budget preparation. Allowances can be a single price (such as "all plumbing fixtures and hardware, total allowance included,

FIGURE 5.1 Entering Formulas

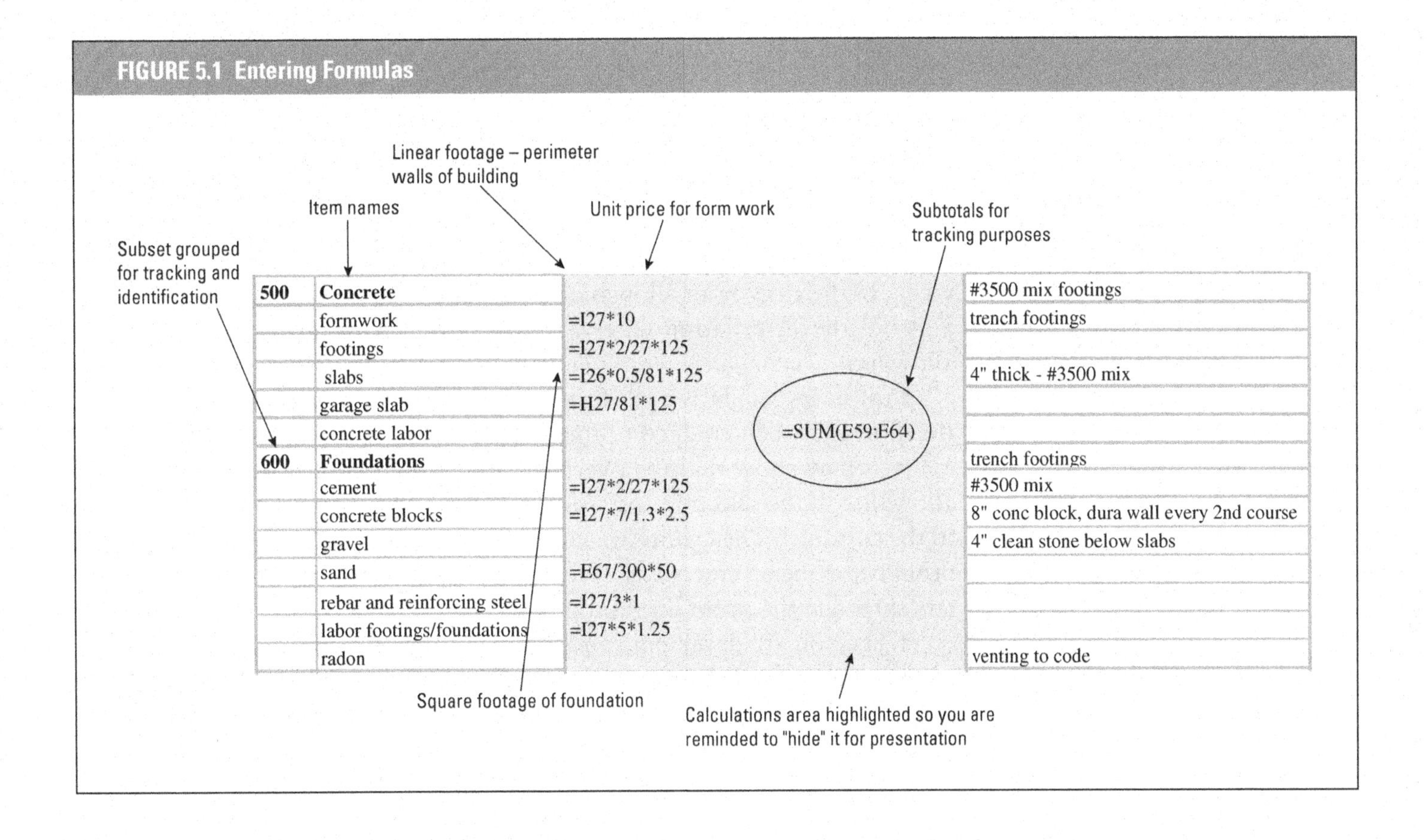

500	**Concrete**		#3500 mix footings
	formwork	=I27*10	trench footings
	footings	=I27*2/27*125	
	slabs	=I26*0.5/81*125	4" thick - #3500 mix
	garage slab	=H27/81*125	
	concrete labor		
600	**Foundations**		trench footings
	cement	=I27*2/27*125	#3500 mix
	concrete blocks	=I27*7/1.3*2.5	8" conc block, dura wall every 2nd course
	gravel		4" clean stone below slabs
	sand	=E67/300*50	
	rebar and reinforcing steel	=I27/3*1	
	labor footings/foundations	=I27*5*1.25	
	radon		venting to code

$4,000") or a unit price (such as "tile allowance of $4.00 per square foot included"). Both of these descriptions cap the cost to the contract and pass overruns to the customer rather than eroding profits.

When you include a single-price allowance for an assembly of tasks, you describe the scope of the materials in detail. When allowing $300 for a range hood, does that include wiring or ducting? Is the kitchen sink included with the plumbing fixtures, or is it a part of the kitchen cabinetry line item? Should the sinks be part of the countertop price if the selected tops are integral-top-and-bowl (ITB) units? You will need to address these and other questions prior to signing a contract and moving to construction. Crisp wording can deflect later conflicts.

When you use an allowance for materials, you must define the installation and whether the allowance is materials only or an installed cost. This difference can be a critical element. A tile installation in diagonals in a specific pattern, possibly set with a certain adhesive method or subject to a lot of cutting and fitting can surely raise the installation price. Here the language could read, "Tile materials to be selected by owners; $4.00 per square foot allowance for materials cost; standard pattern installation to be in set w/ mastic and grouted with standard grout; installation per manufacturers specifications." The rationale for this installation language is to prevent unbillable cost overruns from the tradesperson accommodating the owner's requests for complicated patterns.

Another way to handle the possible divergence of materials and installation costs would be to use the word *allowance* to include both, such as "allowance for cabinetry—$12,000 installed." This positioning allows the estimator to move past complicated installation without worry to profit. The selected cabinets may occasionally require a lot of cutting and fitting, custom crown moldings, or shimming to old floors and walls, each of which could add days of additional installation over simply screwing stock boxes to the walls. In the case of an allowance that includes installation, you can track and bill the installation costs as well as the materials cost against the allowance.

Customers regularly ask: if the materials cost less than the allowance, is the difference then credited to them. The direct answer is yes. Some further analysis may show that the overhead and profit charged was based on an allowance. If the selection costs less, credit the difference from the allowance to the customer. The credits are not likely to be marked up for overhead and profit before being entered as a credit on a statement, which leaves a few dollars spread in the profit line untouched.

When you are developing an estimate, using allowances allows progress to continue when some materials are not selected and designs are not complete. The core role of an estimate (besides protecting profits) is to get a contract signed that provides a budgetary comfort zone for the client. Using allowances permits this contract commitment well before all the selections are complete and gets the work chronology started.

Linear Footage

The use of linear-footage pricing in an estimate is most appropriate when you purchase the materials or labor as linear-footage items. In this way, you align the costs to the numbers in the estimate. You can do the takeoff from the plans with a scale or by tape measure or measuring wheel during a site visit. A potential pitfall to profits can occur when you enter a linear-footage estimate and then buy the component by square footage or worse day-rate labor and unknown materials.

We have a fixed quote from a trade contractor for linear footage of gutters and leaders and unit costs for elbows and accessories. When populating the estimate cell for gutters, we simply look at the elevations with a scale or a drainage plan if one exists. We enter the number of linear feet in one cell of the spreadsheet, which multiplies it by our contract cost for gutters. Thus, we obtain a solid number to work with for the contract. If we were to send in our own carpenters to do this installation, we don't control the time, and we do not have a gutter-forming machine, so handling and also materials pick-up become uncontrolled costs.

Uncontrolled costs are a risk to profit, so in the gutter example, the right number comes from a linear-footage calculation that you can back up with a purchase price that matches the estimate. The industry standard for how an item is purchased provides a driving force for how you should estimate it.

You will find more on estimating production in Chapter 9, but for now, remember that the method for adding the cost of a selection to the estimate should reflect the methods by which you can purchase that item.

Several marketplace advocates will build "assemblies" into a unit cost and enter that number in an estimate. An example would be to estimate the linear footage of interior walls. This assembly would include framing, drywall, trim, finishing, and maybe even electrical outlets prorated to one per 12 linear feet to meet the electrical code. These assembly calculations should result from close inspection and dissection of all the component costs before an estimator can go back and bulk a number of components into one unit cost.

The immediate risk to profits is that you can't buy one foot of that assembly. In fact, you can't buy any amount of those assemblies because they are not sold as such. The difficulty extends beyond purchasing; it also involves tracking costs. If you were to estimate linear footage of 8-foot high interior walls at $50 a linear foot and had a 10-foot wall, the estimated cost would be $500. You would be unlikely to find a framer, drywall installer; trimmer, and cleanup person to do the job for anywhere near this amount. So the solution is to have one person, a lead carpenter, do all these tasks, which makes plenty of sense. The person who can do it all gets the job done on time and budget.

Figure 5.2 compares the pricing of assemblies to the pricing of a day rate for a tradesperson. On the small jobs, the day rate is a higher estimated cost than the assembly estimate. On the large job, the cost estimate for the assembly is higher than the day rate. One analysis could be that when you estimate small jobs at linear-footage-assembly prices, they may not create profits. However, if you estimate them at a day rate, you get a higher-priced estimate. An industry peer looked at job costing over time and found his most profitable jobs were those lasting no more than a few days. He billed on a day rate and got a hefty markup over cost. His results were consistent with the chart in Figure 5.2 that shows you gain an advantage when you estimate and bill small jobs at the day rate. You will find plenty of variables in this example because it is for illustrative purposes only.

Linear-footage units have their place in an estimate, but you need to take great care if you use them for tasks that involve multiple activities. When you calculate the linear footage of a retaining wall, height, soil conditions, backfill specifications, or drainage design all could be wide variables. They may introduce additional costs not reflected in the underlying formulas that create a linear-footage price. Speaking of the linear-footage cost to build a road in a subdivision may make for interesting cocktail chatter among builders, but until you buy a road by the linear foot, the possible risk to pricing is high.

FIGURE 5.2 Day-Rate Cost Estimates Versus Assembly Cost Estimates

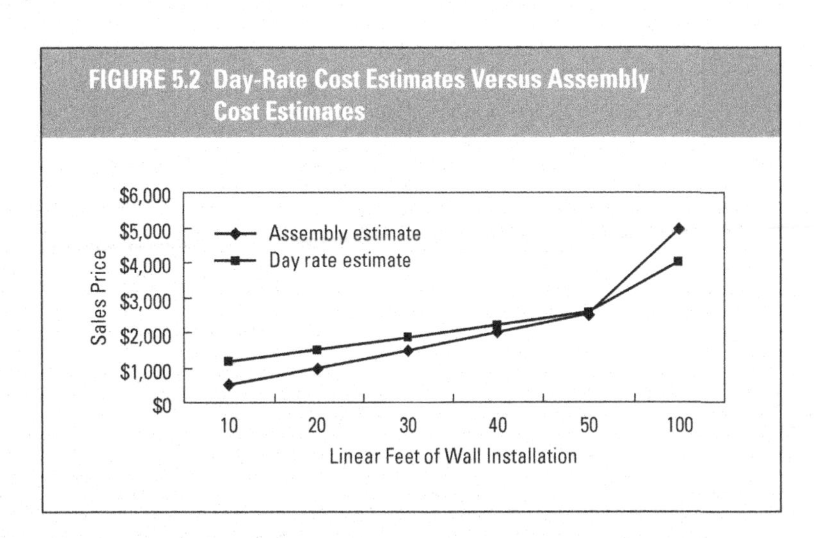

Square Footage

The use of the square-footage estimating system is similar to use of the linear-footage estimating system. You can best use it when you can buy the underlying item by the square foot. This parallelism keeps the estimate aligned with the purchased units and enhances price predictability.

You can increase the calculated footage for waste (hardwood, insulation) and materials loss resulting from layout (carpet, sheathing). Calculating footage is worthy of a book in itself; in fact several on the market use many definitions that include rentable space, usable space, net space, gross space, and so on. The purpose of square-footage calculations can vary as well, from materials estimating, leasing, and sales to building design and zoning compliance.

Estimating is a process, not a math formula for a hard and fast result, so you should treat every discrepancy you identify as an opportunity to enhance the process, not as an error.

A heating design will use one number for the interior space, and the hardwood installer is likely to find a different way to calculate the number eliminating stairwells and the like. A roofer may round up the number to account for hips and valleys when calculating shingle orders or use a formula for outside wall sizes times the pitch of the roof while another may count the sheets of exposed sheathing.

One way you can monitor line items for square-footage accuracy over time is to track the number from estimate to final invoice and identify discrepancies. As you find these discrepancies, you can update your database, implement clauses for additional billing opportunities when the sizes are off, or tighten your trade contractor agreements to align costs with the estimate.

If you go back to the flow chart in Figure 1.2 in Chapter 1, you will see these alterations to the process on the bottom of the chart as feedback into the database. The square titled "Estimates vs. Actual Report" with an arrow leading to the circle titled "Revise Database" illustrates this process review.

Contract Prices

A contract price is a specific agreement to supply a certain amount of materials or perform a fixed amount of work for a defined price. You certainly could create a contract agreement (or purchase order) for every line item of building activity. Theoretically, this procedure would pass the risk to all of the other contract parties and minimize risk to the builder or remodeler of having the accuracy of estimates vaporize. The high-risk components of a contract price include the scope of work, materials specifications, and performance standards.

Scope of Work

As to the risk in scope of work, you need to define its start and completion as well as the breadth of the work. The dark pockets of activity that are not well defined are often overlooked, so you need to address every item. A linoleum installer may have a good contract price for the materials but then not find the subflooring installation acceptable to put down the product. The installer may proceed to refasten the floor, dash patch the joints, and bill for some amount of additional labor for the preparation. These costs could tip the line item over and above the estimated costs. The way to prevent those overruns might be to include in the flooring contract that the installer is to look at the site before installation and notify the builder that he accepts it "as is" or report modifications that may be needed. At the minimum a step in the process now includes verification and acceptance of the site, which are often points of conflict.

Does the contracted item include debris removal of excess scrap or at least putting it in an on-site dumpster? This activity alone could mean you need to schedule a regular cleanup between trades. This necessity commonly occurs and you probably need to insert a price in the estimate for regular cleanup beyond the final cleaning of the site for closing or final payment. You can define the risk transference to the costs of cleanup in the trade contract language. Several excellent sample agreements are available in publications found at www.BuilderBooks.com. A sample of such contract price protection follows:

Section X. Cleanup

X.1. The Trade Contractor will be responsible for cleaning up his or her work on the job on a daily basis, including all generated construction debris, drink cans, food wrappers, and/or other trash. If necessary, the [builder or remodeler] will back charge the trade contractor for the appropriate cleaning trip by deducting cleanup costs from progress payments.

Contract prices and performance should also include stipulations for insurance coverage, change orders, payment schedules, warranty, and materials specifications and substitutions.

Materials Specifications

The second area of risk in contract pricing is the materials specifications. You need to align these specifications with the customer's contract. Selling one material to the customer and buying a different material from a trade contractor is surely a formula for a later conflict.

We sold a flat roof replacement as a "rubber roof" only to have the roofing contractor suggest a different material that would cost no more and was to be "a better job." In deference to his experience, I let him install it without changing the customer contract. A delinquent final payment for a "better" roof ended up in court only to have an expert identify that the product installed was not per the contract specifications. The flaw was not that the

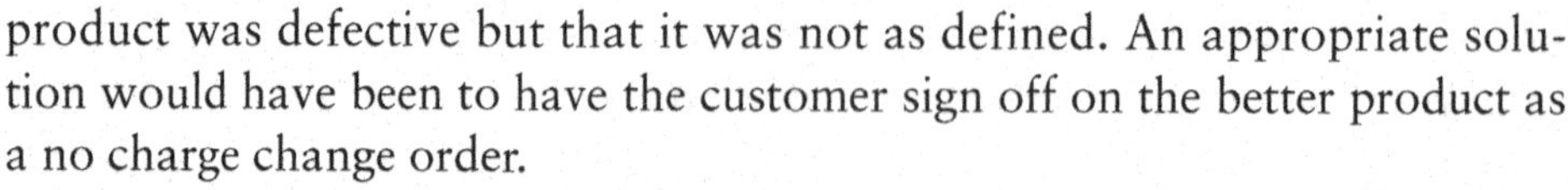

product was defective but that it was not as defined. An appropriate solution would have been to have the customer sign off on the better product as a no charge change order.

Alignment of contract specifications sold and contract specifications purchased must match. Otherwise, the customer must sign and pay for a change order before you substitute an upgrade for material not covered by the contract. Beyond alignment, you need to get a warranty on the materials consistent with a builder's or remodeler's obligation to cover that product in the home or a remodeling job. A handy way to handle generic specifications in materials selection is to get a brand and model into the contract that a local supplier can provide. Rather than describing a high-hat recessed lighting fixture, the line might say, "Halo H-7 housing with a #310 trim." This description sets a benchmark for quality and cost that narrows risk to product selections.

> We regularly have customers make selections that are substantially more expensive components, and these cost increases need to be reined in with change orders describing the cost and the upgrade.

Performance Criteria or Guidelines

The third risk to contract pricing is the performance criteria or guidelines by which the work is judged. These guidelines can be defined by law (start and completion date for performance), building codes and mechanical subsections (design criteria), or what is vaguely defined as "workmanlike manner." You need to create and communicate in writing some understanding about the performance guidelines for the contracted work.

Most likely you would develop these guidelines in writing over a series of jobs. The failure to define the quality of work expected falls directly on the builder or remodeler. When price is the criteria for entering into a trade contract, you must also keep the level of performance tightly defined.

Day Rate

You can input units of day-rate pricing into a spreadsheet cell when you purchase the item at a day rate. Look at the example of backfilling a foundation. Unless the builder or remodeler owns the equipment and has already moved it to the site, backfilling a foundation is likely to involve the excavator moving a piece of equipment to and from the jobsite and completing the work while the equipment is on the site. The excavator probably would bill this job at a day rate for a small-track excavator of $600 per day. The day rate for this equipment would include fuel, machine, operator, and insurance. For estimating purposes, you would list this backfilling cost of $600 in the spreadsheet.

A sharp builder may be able to see that the backfilling only takes half the day so the builder might use the rest of the purchased day to create swales, pack a dumpster, pull some stumps, or keep the driveway grades intact. You can closely manage day-rate costs to get additional productivity. Similarly, you also must closely manage day-rate costs so that the work doesn't run on endlessly without accountability.

Quantity Takeoff

A quantity takeoff in an estimate cell is actually counting and pricing a certain number of identifiable items. For example, counting and specifying the windows going onto the jobsite is a relatively easy task demonstrating quantity takeoff. The plans show the windows with sizes and possibly a manufacturer recommended by the architect or owner. The builder or remodeler can drive this brand decision to a favorite manufacturer. The builder or remodeler sends the quantity takeoff to a supplier for a hard current quote. Use fax software or cut and paste details from the spreadsheet into an e-mail to get the supplier involved in building accuracy into the estimate. Spreadsheet capabilities include selecting certain cells and printing that selection (as a fax) for the salesperson to quote (Figure 5.3). The notes column in the spreadsheet can help define the quantity and items included in the request for the item's cost estimate (Figure 5.4).

Remember estimating is about minimizing risk to profit. It is not about making friends or seeing how cheaply you can build a project to create a low bid. All steps to risk reduction help to ensure that your business succeeds for years to come.

The seven ways to get the numbers into the spreadsheet also apply to the existence of items in the spreadsheet. The first question is: Are you buying the item the same way you are estimating it? If the answer is yes, the pricing should be aligned and accurate. If you are not buying the item the same way you are estimating it, you have two possible solutions: Begin to estimate the way you buy, or buy the way you estimate. You do not need to have a line item under drywall for the number of drywall adhesive tubes required in the house if the drywall contractor is giving you a flat price contract that includes materials. This level of detail is not how you buy the work so the estimate should mirror your buying profile.

Disconnected methods are risky to profits. A lump-sum estimate for plumbing labor is not appropriate if you are paying the plumber a day (or hourly) rate. A less-risky arrangement would be for you and the plumber to use a fixed-price contract for specified work. The builder or remodeler is taking all the risk if the consumer contract is a lump sum and the plumber bills whatever days he or she works on the site without a cap to the costs.

The quality of the estimate is only as good as the quality of the plans and data collection. If you go back to the original diagram of the estimating process in Figure 1.2, "Work Flow in Estimating," in Chapter 1, the earliest steps include data gathering. You add risk if your communication—through the contract, its addendum, plans, verbal or written instructions, or level of detail in the specifications—is not fully up to construction communication guidelines. You need to make these details clear, unambiguous, and accurate. Only then will the seven ways to get the numbers reflect what is going into the project and align estimated costs with purchase methods.

FIGURE 5.3 Quantity Takeoff for Windows

Please quote the following windows (VAN LOAD)
Brand ______ Series ________, white, w/ screens

Qty	Window
1	3046
4	3046-2 (narrow mull)
3	21032
1	21032-2 (narrow mull)
5	26310
2	2642-3 (narrow mull)
1	DHT 2421
1	2836-2 (narrow mull)
2	2836
2	1832

FIGURE 5.4 Takeoff Entries

Estimate Spreadsheet			
Items	**Quantity**	**Pricing Type**	**Specifications/ Notes**
		Lump-sum pricing	
		Allowances	
		Linear footage	
		Square footage	
		Contract prices	
		Day rate	
		Quantity takeoff	

CHAPTER 6

Using Retail Pricing at Every Line

The best-case cost inserted into a line item in an estimate leaves nowhere to go but up. If you are holding only the potential to slip backwards, you are not in a good starting position. If you use this lowest cost, what is the likelihood of getting best-cost materials, best-cost production, and error-free output? The likelihood of being able to reproduce a best cost in the field for every line item in the estimate is indeed remote.

Put the Risk into the Line Item Not the Bottom Line

During the preparation of the estimate, you can place costs in the line items that are your best-case numbers for a component of the work, or you can embed in it a "street price" cost. This cost is derived as if you need a new, reputable trade contractor to provide a subcontracted price for the job. If you were to hit the streets looking for this reputable trade contractor to do this work, what would that person charge? The definition of *street price* is the going rate for reputable people to do fine work.

When preparing an estimate, you need to use the street value of the work rather than your best-cost number. This practice allows your production crew the flexibility to either (a) subcontract the component to an established trade contractor (who ideally is charging you less because of your long-standing relationship), or (b) do the work yourself with your own employed labor. When you are using best-possible line item prices, any slight error triggers a cost overrun from the line item, and the cost overrun immediately tarnishes profits. Street prices create opportunities for additional profit at each line item if you can produce the work at less cost than the market offers.

For example, with a phone call to several roofers in the area, you might find the cost of installing roof shingles is $65 per square for installation only. You may have a long-term trade contractor who does this work for you for $50 per square. What number should go in the estimate? Is it the $50 you are paying based upon your long-term relationship, or the $65 you would spend if you called a fresh contractor from the bullpen? The best answer is $65 per square. The work is worth $65 to anyone calling for prices and planning to hire from the available pool of installers. The discount of $15 per square is yours as compensation for your investment in the relationship. The benefits to your trade contracts include that you

- supply a continual stream of work to this trade partner
- pay your bills to them on time
- have checked their insurance
- monitor their safety

We have developed a pool of reliable, reputable trade contractors over time who consistently deliver high-quality work and materials. Their cost proposals are in our estimates. These costs do not fluctuate unless the customer changes the scope of work or substitutions require a change order to the customer. Hence, our risk to profits drops with the consistency of this teamwork.

Your discount compensates you for your work in relationship building. You should be paid for this relationship building and for the work of obtaining preferential pricing. Were you to use the actual price of $50 per square, your client benefits from the $15 discount off the market rate. You are in not in business to make money for clients; you are in business to make money for your company. The value you create for your clients is in getting the job done well, supervising the work, and expediting completion by getting them a better job in a tighter time frame with less risk or fewer headaches for them.

The graphic in Figure 6.1 demonstrates the path used to get a street price into the estimate. Note that the path does not go through uninsured, unlicensed, or unqualified tradespeople. Their cost may be lower, but the risk to profits is much higher.

FIGURE 6.1 The Path to a Street Price for Roofing

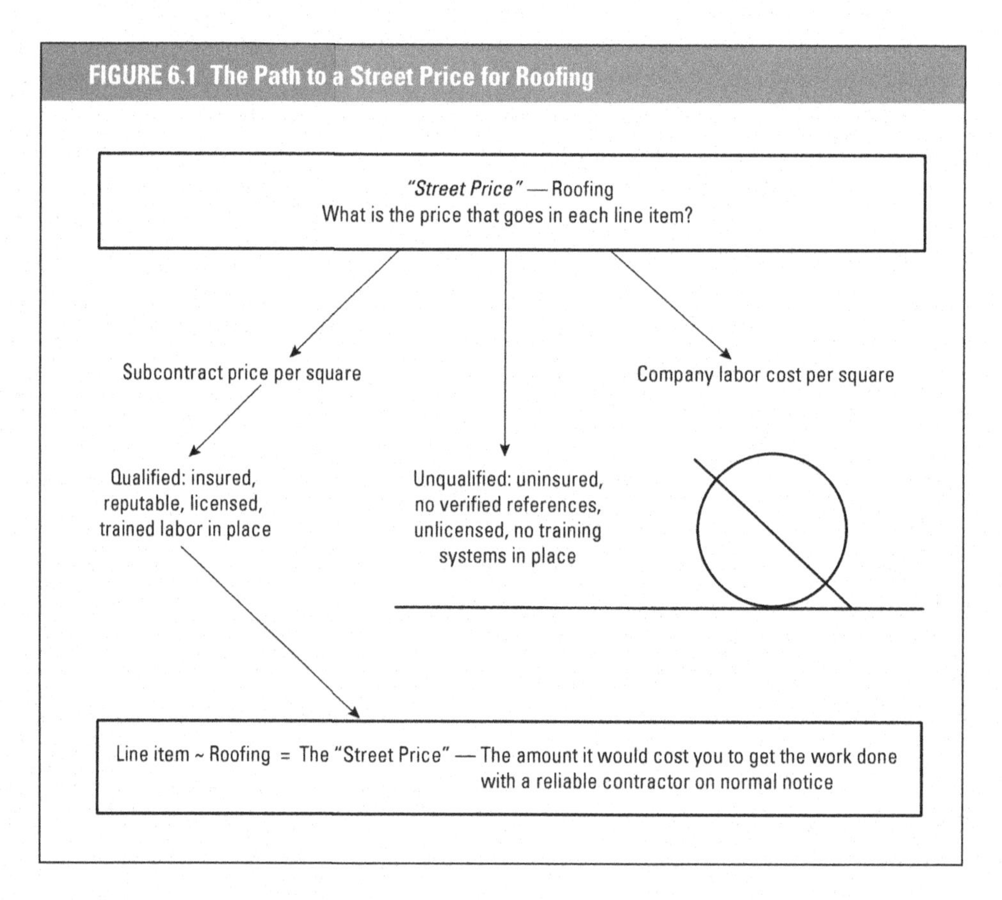

Often a builder or remodeler in a bidding situation wants to lead with his or her best-cost numbers, and the first items to go awry during production erode profits. Using all best-cost numbers within the estimate, and the likelihood of consistently producing at these numbers is near foolhardy. Reminder: The reason to seek and get the work is to create profits—not simply to get the work. When Stan Ehrlich, formerly of Depot Homes, Inc., in Clinton, New Jersey, discusses a cut-throat price offering, he advises: "There's bad work and there's no work, and no work is better." The definition of the term *good work* is profitable. Without the profits on the bottom line, you have no reason to take the risks in a contract. Submitting a high-risk, low-profit estimate can only lead to bad work. His firm regularly rejects work that is not going to add to the company's annual profits.

Nullify the Risk at First Entry

Using street pricing at every line item opens opportunity for additional profit within your estimate. If you can do the work at a lower cost, you can make money on the line items in addition to the projected profit found on the bottom line. This street pricing allows you to do the work at a possible lower cost with in-house labor. If you can do so, the additional cost savings go into your bottom line. In effect, your firm is acting as a lower-cost trade contractor for your own job (Figure 6.2).

As Chapter 4, "Spreadsheet Estimating," discussed, you should have a line item for the supervision aspect of a building contract. What would super-

FIGURE 6.2 Pricing Strategy: Beating the Estimate at the Line Item Level

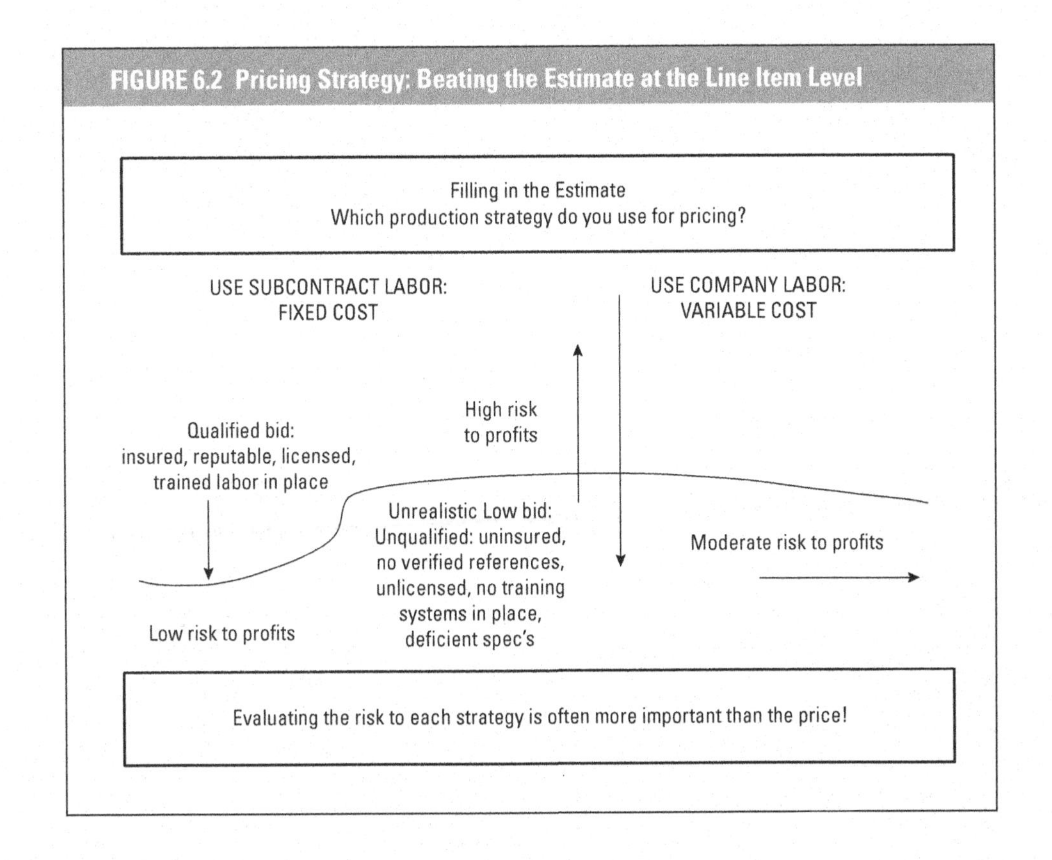

vision of the job cost? What would you need to pay for a competent individual to watch over the job? That cost should be in the estimate. You need to decide whether the job would justify a full-time supervisory position (or even several positions). Can the supervisory job be done in-house in a few hours a week? The cost of supervision is real and billable within the estimate. A common sense approach to managing this supervision line item is to empower employees already on the job to multitask this responsibility. This management technique is commonly called a lead-carpenter system. The estimate should reflect both skills. You need to accommodate the cost for both the supervisory and the carpenter roles within the estimate.

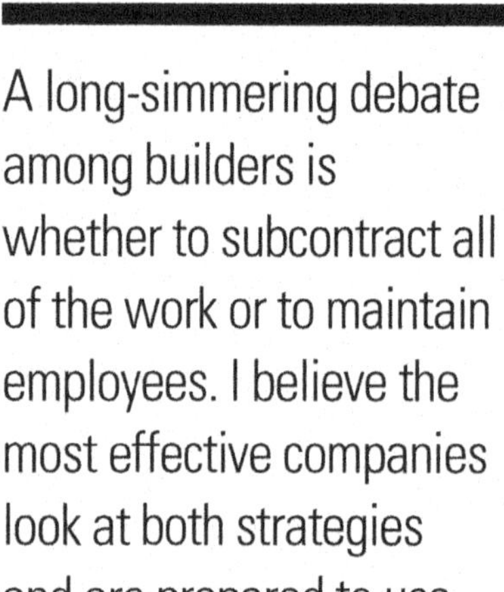

If you choose to pick a few line items and do the work in-house, you should be paid the full retail value of this work and add it to the bottom-line profits as you hammer down your hard costs.

If you can be cost effective and efficient with labor line items completed by employees of your company on your own job, go that route. Several rules of thumb govern this decision of whether to subcontract a particular piece or phase of the work or do it in-house. In roofing, the rule of sixes governs: For more than six squares or more than a 6/12 pitch, you would call in experienced roofers that clearly are faster than in-house labor with sporadic involvement on the roof surface.

For the tiniest of jobs (less than 6 squares), the overhead cost of mobilizing outside help and coordinating schedules is often more burdensome than just doing the work and putting the line item behind you. Often short-term tasks involve a minimum cost to complete so the unit costs rise quickly. At these higher unit cost rates in-house labor can easily compete, better the street price, and drop additional profits into the bottom line. The goal is to make money at every line item in the estimate.

An example of pricing for labor in an estimate should offer a clear demonstration of the sharp differences in applying this strategy. If a company is paying an employee $25 an hour with an overhead burden of 35%, the cost to the firm is $33.75 an hour. The replacement cost of this labor on the street is likely to be $60 per hour for a short-term, highly skilled mechanic. Which price goes in the estimate: $33.75 an hour or $60 an hour? The answer is the street price of $60 an hour. If you need the labor to complete the work, $60 is the value of the carpenter's time. By choosing to take the responsibility for this full-time employment and cover the benefits and overhead, this difference is yours as compensation for those efforts. Alternately, the carpenter could be working elsewhere for $60 an hour and billing at full retail. Either way the work is worth $60 an hour and should be included in the estimate at that rate.

What is the value of putting heavy equipment on a jobsite? If you own it, your out-of-pocket costs may drop, but keeping the piece of equipment in working order and delivering it to the jobsite ready to work anytime bears a retail price. You include that retail price in the estimate. Should you choose to invest in heavy equipment, do so for convenience, access, and reliability—not to lower the cost to your customers.

You may find another example of street pricing in the lumber line item. Hopefully your long-standing lumberyard is giving you preferential pricing and service that you can convert to profits. When you use your best-cost pricing and preliminary material takeoffs, your bottom line pays for every stud you omit. Alternatively, the lumber list should be slightly "soft" to accommodate a few boards to use for a scaffold, for missed cuts, or to replace those boards that are twisted and in need of replacement. By slightly "soft," I mean that you have a few extras built into it. The shortfall should be paid from softness in the line item rather than the profit line.

One of the best uses of the supply chain beyond getting the lowest price involves engaging the supply chain staff as a team member, mining the firm, and substituting its services for your otherwise hard costs. Mining the services is simply understanding what the supply chain can offer and using these services. This could be a materials delivery that is short a couple of items and the vendor will make the delivery; a liberal return policy that allows you to convert some overage back to a return ticket; timely deliveries to minimize onsite labor losses; and credit terms that enhance your cash flow. You can easily convert these benefits into cash for the bottom line. Builders seldom mine these value enhancements for the bottom line.

Mining trade contractors' benefits on the site may be as simple as having them clean up after themselves to getting a discount for prompt payments. As long as street pricing is going into the line item costs, the opportunities to drive down expenses and convert the savings to profits are all through the estimate.

Early Learning Strategies: Calculate the Actual Numbers at Each Line Item

Early learning strategies for getting these street prices include getting bids from qualified trade contractors. Every job is different so you could get a lifetime of proposals for a series of specific jobs. Some estimators will not put a cost in a proposal unless backed up with a hard quote from a supplier or tradesmen. This policy is clearly a strategy for reduction of risk to profits. Gathering these numbers does take time and can raise the overhead to your trade contractors as you challenge them to go look at every job.

Early in the mastery of estimating, this strategy can give you a feel for the numbers, help you build a database, or let you make some qualified guesses for use in prequalifying customers. As your estimating comfort grows, you are likely to decrease your dependence on precontract pricing and replace it with a series of shortcuts, rules of thumb, and allowances that speed the delivery of the estimate for customer approval while minimizing risk to profits.

A careful reading of the subcontract bid specifications will help you to (a) eliminate duplication of costs if two trades propose to do overlapping work and (b) plug holes resulting from omissions. Because your estimating spreadsheet is chronological, every step toward completion should have a retail number in place.

A plan review or jobsite inspection with an experienced estimator or trade contractor is invaluable. They can help identify conditions that could be costly to build or work around. For any of these unique conditions, you need to have a priced solution or a strategy for risk reduction. Chapter 9, "Production Tracking," details a number of these risk-reduction clauses. These site visits with experienced tradespeople are a fertile learning ground for what to look for on the jobsite. Some site conditions may require actual testing (soil compaction, presence of radon, presence of hazardous materials). You can best accommodate testing with a disclaimer or an allowance that covers costs or you can redirect the cost back to the owner.

Our excavation proposals include a number of disclaimers for soils conditions and the like. We then forward the disclaimers to the consumer in an addendum to our estimate so we don't get caught with a cost overrun that is not reimbursable.

An organized trade contractor should be able to give you some unit costs for baseline estimating. Examples to use as estimators' crutches (shortcuts) may include a price per square foot for framing, a linear footage price for gutters, or a square footage price for painting. Each of these trade contractors lowers your risk of profit erosion by providing a fixed price that transfers the risk and responsibility for completion of that line item to them.

Using street pricing puts the limited risks of production or materials purchases into the line item and not in your bottom line profit. You have now added another layer of protection and defense to the projected profit number.

CHAPTER 7

Minimizing the Workload

Prequalify which estimates to tackle. Estimating is a drain on overhead, and the less you have to do while keeping up sales and production, the lower your overhead costs will be. In a working year of 50 40-hour weeks, you have 2,000 hours a year to spend generating value for others. Maybe the committed entrepreneur works 2,500 hours a year, and the obsessed claim 3,000 hours a year. But sooner or later you simply run out of time. As discussed earlier, speed and accuracy in estimating are inversely proportional. The faster you go, the less accurate the estimate will be and the more profit you leave at risk.

Some people in the industry would suggest that doing estimates faster would solve the problem of the consumption of time required to put out a large number of proposals. This solution may yield a long list of inaccurate estimates that challenge profits. Unless you have a narrow and well-defined product line or service item, a complete estimate is likely to take a serious commitment of hours for you to complete it.

I suggest simply doing fewer estimates.

If you are selling only garage doors, you may have 30 styles and a handful of sizes, each with a price ready to go to market. If a customer calls, all you need to do is match the customer's selection to the opening size, maybe verify the size with a field inspection, and send out the number. To complete an expedited estimate for building a house is nearly impossible. Several thousand products go into a home, and possibly 35–50 trade contractors could be working on the jobsite through construction. Shaving the estimating time is nearly impossible because of the mandatory phases of data collection, assembly, hand-off to production, and tracking.

You can use estimating as a sales tool to prequalify customers based on the client's ability to afford the work. An easy question to ask the potential client is, "What is your budget for this work?"

One technique to elicit a potential client's budget is to walk the client to an extraordinary price laden with options, upgrades, and features and ask them if that is what they had in mind. Most potential clients would cringe and admit that it is not what they had in mind. You can also use a short guessing game, and it may get to a budget number. A fair game is to ask if the budget is $50,000 or $150,000. Many will nod to a range of price.

You then need to establish whether they can afford the scope of work you are proposing. You may be able to downsize the scope to get to the budget or

I have been told everything from, "You're the expert. How would I know?" to "We have no idea what the budget should be."

FIGURE 7.1 Some Sales Criteria Limit the Jobs You Take

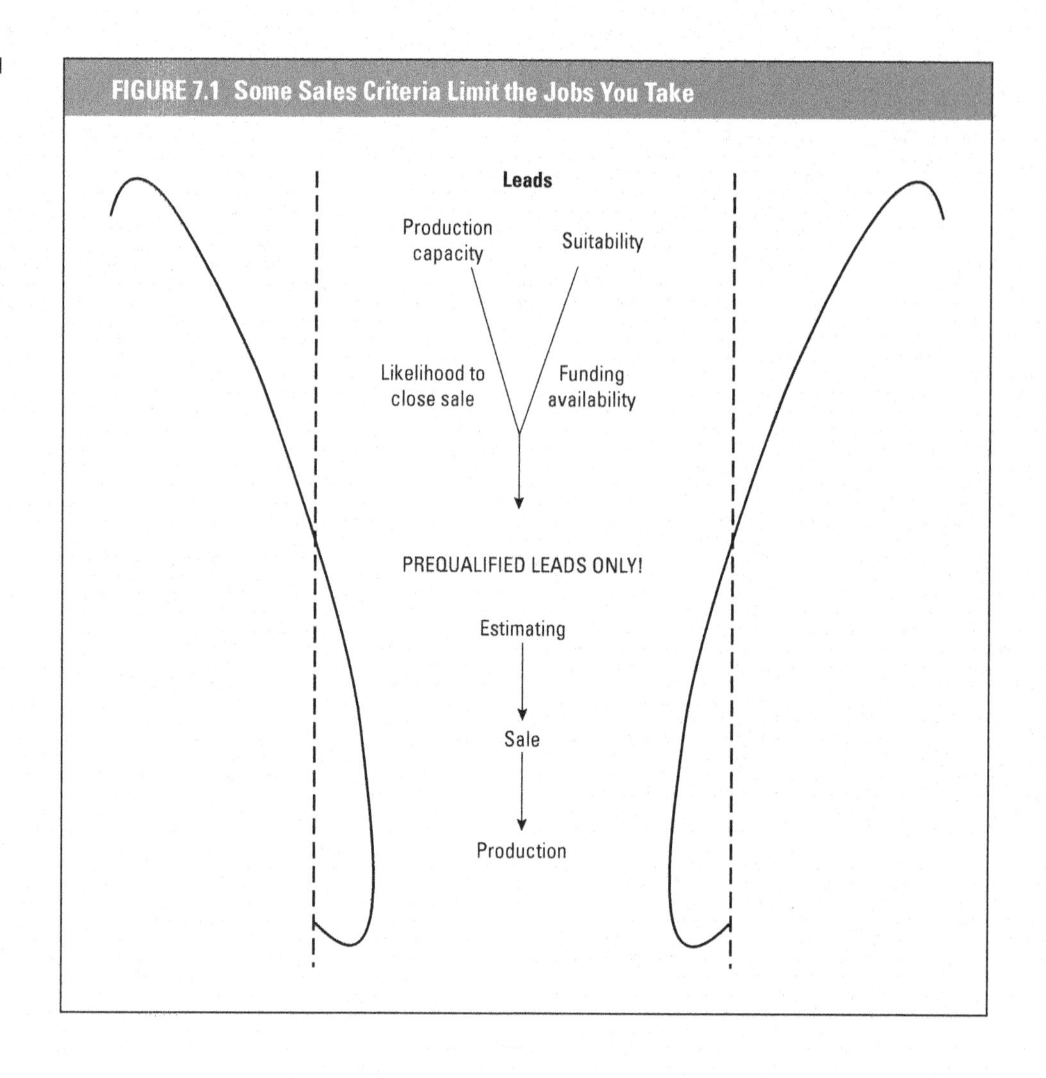

We can animate customer service for personal satisfaction, speak of our role in housing as fulfillment of social responsibility, or glow about the joys of entrepreneurship, but without deposit slips, the business goes away.

get financing help to broach affordability. Not doing a fruitless estimate saves time, and it minimizes overhead wasted in unproductive customer calls and proposals. This prequalification of clients is mandatory for controlling your time.

Some builders and remodelers describe the prequalification process as a funnel. The wide mouth at the top represents the leads in your market. It tapers as you winnow leads to estimates produced and further tightens around those clients who sign contracts that allow you to make out a deposit slip (Figure 7.1). Business is about filling in deposits slips.

You need to eliminate useless leads through prequalification. Some of the criteria could be

- the type of work
- travel distance

- budget constraints
- financing or payment terms constraints
- schedule
- availability of manpower
- the number of builders the client is talking to
- lack of preplanning on the customer's part

You can use a number of these prequalifying techniques to winnow leads that take far less work than actual estimate preparation. A fair series of questions could include the client's budget, the prospective client's time frame needs for construction, the suitability of the work for your company, geography, and the builder selection criteria the prospective client used.

When asked if I am interested in looking at a job, I ask the prospect directly, "What are your builder selection criteria?" The answers are telling. If the person stammers or hesitates, the stammering or hesitation probably results from an inability to admit that he or she is looking at price. Price is typically in the consumer's crosshairs. I tactfully remind the person that the lowest bid is often created by the estimator with the most mistakes in the estimate. I suggest you add this question to your prospective customer interview. The answers range from "we want you" (great answer) to "we are gathering bids" (wrong answer).

Most sales cannot be completed without meeting the customer, probably spending several hours of data collection on the site with the customer, followed by several hours of preparing the estimate, and then back for personal delivery. You can establish several phone techniques and prequalifying criteria to do the early screening. Given the described process of several hours of data collection, several more of office time, and a presentation meeting, the builder's time burns up quickly. If you assigned an hourly value of $75–$100 per hour to the investment in the lead before conversion, the cost in overhead to deliver a proposal is likely to run $750–$1,000. At that rate, you could cull the lead in a 15-minute call, and your rate of income per hour would be 4 times the investment to get to the estimate, or $4,000 per hour.

Clearly the best thing you can do is to query and dismiss unproductive leads as fast as possible. Securing a contract to build or remodel is a two-way commitment. Not only must you morph the customers' needs into fulfillment, but they also should be an ideal fit for your company. A mismatch either way can risk your profit. You should spend some time in a business planning exercise to define what your ideal customers look like so that when they come to you, you can screen effectively to get the right ones quickly (Figure 7.2).

The fewer estimates you have to do, the lighter the workload. Prequalifying leads is far more effective than immediately agreeing to a site visit or plan review and tying up multiple hours in overhead.

Charging for Estimates

Builders and remodelers continue to have an ongoing discussion in the industry as to whether to charge for estimates. Surely, if they are using you to

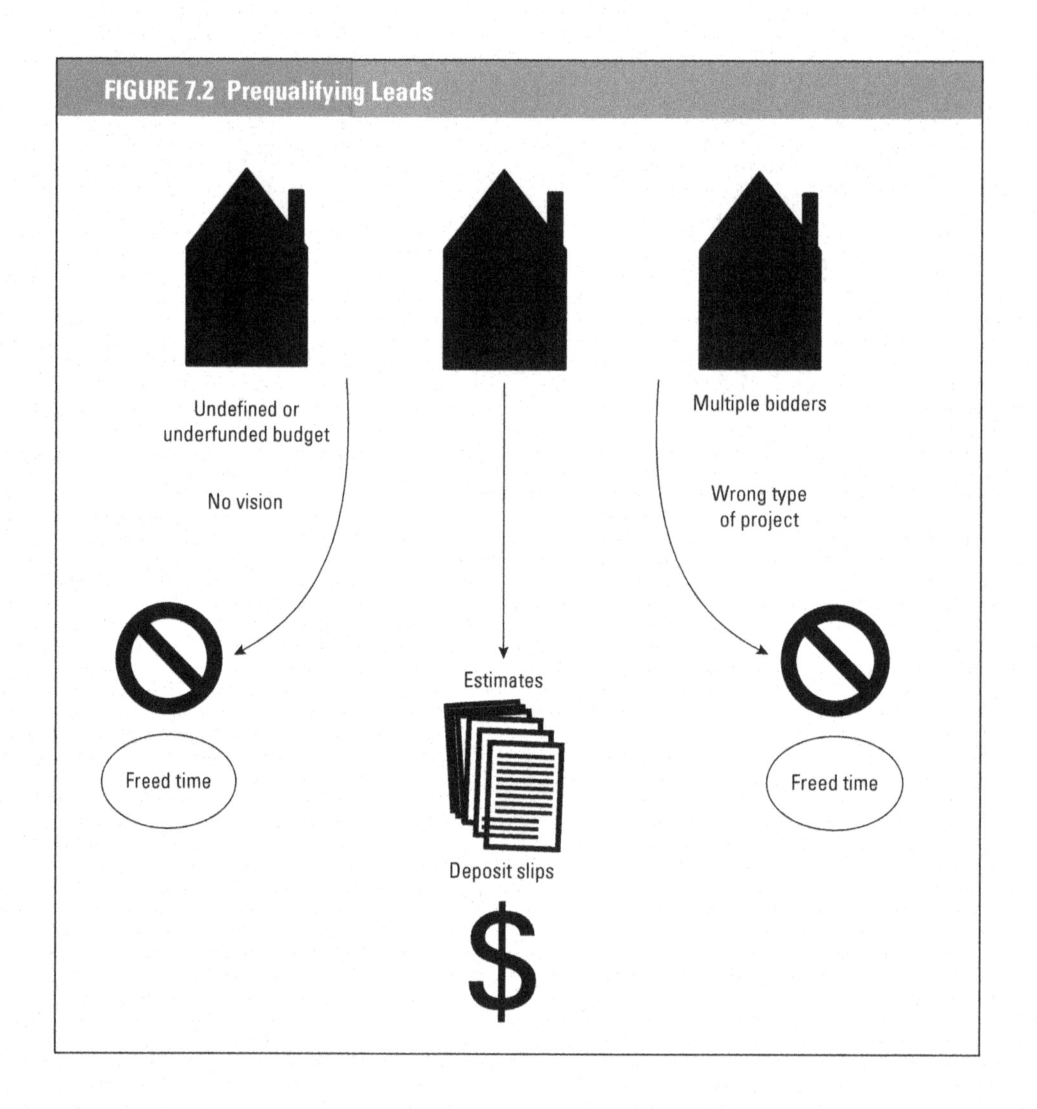

FIGURE 7.2 Prequalifying Leads

collect on an insurance claim or to establish the value of defects in a home after an inspection, these events should be billable. Your role is as an expert or consultant on construction value, and the scope of work is to be the expert, not the contractor. Similarly, as an expert witness in construction defect claims, your time should be billable.

In design/build work, you provide a valuable service in developing the design, understanding the client's needs, and proposing a range of solutions. You need to bill for each of these separate services. You can use the cost of estimate preparation as another prequalifier of the customer's interest. If a prospective client is willing to invest a few hundred dollars or more for you to develop a price for his or her vision, that prospect is likely to be a highly motivated buyer.

The estimating step in the contracting process can easily be a nasty bottleneck on the way to profits. Logic would suggest that opening that blockage can happen in several ways. Software vendors may suggest that their technology can reduce the time you need to create an estimate. A management consultant may suggest that you compartmentalize your personal time-management system to allow a slot for estimating work.

Mike Turner, a National Remodeler of the Year and voice of Atlanta radio 750 am WSB's *The Home Fix-It Show,* suggests: "Be wary of prospective customer warning phrases. If you hear any of these, you may be headed for the Estimate Money Drain! Think long and hard about knowing when *not* to estimate the work!"

- ☐ "We're not too sure what we want. . . . "
- ☐ "I forget where we got your name. . . . "
- ☐ "We'd like an additional bid."
- ☐ "Price it using the standard stuff."
- ☐ "We're not sure what our budget is...."
- ☐ "We'll figure it out as we go...."
- ☐ "Your price is okay, but we need a breakdown."
- ☐ "Price it with hardwood, vinyl, ceramic tile, and carpet."
- ☐ "We need an estimate for the insurance company."

Or

- ☐ If the customers are arguing over scope or price between themselves
- ☐ If a third party calls you into the process and wants the price, e.g., real estate agent, home inspector, engineer, attorney who does not have authority to sign the contract

The best way to reduce the estimating bottleneck is to simply do fewer estimates. If not, you may be too late in the process to save overhead when you have already invested the time and done the proposal.

In addition to doing fewer estimates, you have a number of other techniques to lower the workload. Discussion of these techniques occurs in the immediately following paragraphs.

Delegate to Others

With some care in counting, someone else can fill in much of the estimate spreadsheet. Certainly an employee can count footage of rooms and perimeter walls. An employee can count the windows and doors right off the plans. You can delegate the title block and entry of the contact information.

This initial plan review for data collection can save several hours of prime estimating time. The economists call this delegation *opportunity cost.* The value of whatever else you can be doing compared to the raw collection of facts from the plan can provide a significantly different value to the company. A second clerical effort might highlight all the steel beams and inventory them for a quote by others. A crisp read of all the specifications with an eye toward the unusual is yet another task for delegation. An employee can add each of the special details from a plan to the spreadsheet as additional rows. Later you or someone else will assign a price to each detail. These unusual items will certainly generate costs.

Use the Supply Chain

Vendors provide a strong source of support in building specifications and pricing. They are an often underutilized partner in the building and remodeling industry. The supply chain brings experience, product knowledge, and troubleshooting help when you need it and ask for it. Some lumber yards will do a materials takeoff, although the framer may be the better candidate for that task. Trade contractors can dissect a plan for their work and quantify prices. On a difficult or unusual project, have the trade contractors join you for a site visit and listen closely to their observations. A little-noticed detail can create a large-cost item or may signal a condition that needs to be brought up to current codes.

Given our company's history of being thorough and delivering a really high-quality product, we are unlikely to be low in any given pool of bids. When price is the criteria or when a cold distribution of architects' plans arrives, these requests generally sit on the bottom of the pile, or we courteously reject them. Hence, a few less leads make it to the estimate process. This reduction in leads helps open the bottleneck in the estimating through sales processes.

Repetition

Creating a repetitive process saves time in preparing estimates. If you follow the same steps each time, the process becomes mechanical. Your process may include gathering some digital photos during the site visit and using them for reference when you are assembling prices. You may use the same trade contractors over and over to gain familiarity with their systems and their output. Many have different styles, so getting comfortable with one may not mean that you can transfer that comfort to the next one.

Save Templates

You can use similar work as a starting point for any estimate. A similar job will have most of the same information and may save a few hours of data entry. A family room of a slightly different size can be used as the basis for the next one to be built. You can construct these templates for baths, kitchens, additions, a whole-house construction, or new homes. Starting with what currently works is a good place to begin an estimate.

Time Management

Focusing on getting the estimate done actually saves time. Maintaining efficiency is difficult if you start and stop, so do a full run-through in one sitting if possible. Take off the items that others need to price and send them out for quotes. A simple note attached asking for a return price by the following week (or day?) may be enough to get all the prices together so your second sitting finishes off the estimate.

Track Turnaround Time

Shawn Draper of IMRE Communications, Baltimore, Maryland, is a marketing wizard. He once told me, "Anything worth doing is worth measuring," and it has stuck. In estimating you need to know how many you close, so you know if your work is effective. Tracking your turnaround time also is important. A simple date-stamp with a pencil when leads come in gives you a baseline to follow and learn how long you take to get the estimate back out again. A prospective customer's tolerance for delivery of proposals follows a logarithmic curve. Depending on the job, that time span could be an hour (current water infiltration) or it could be a month (price an addition).

The customer will let you know when your welcome is running thin. In a professionally run office, you should have some sense of your turnaround time and some idea of where each of the estimates is in the sales process. Lead management software will track this information for you, but you can do it just as easily in a spreadsheet. The spreadsheet tracks all your leads as they mature to estimates and then close as completed jobs.

We have used the same mechanical firm since 1978, the same tile installer since 1984, and so on. Repetition and consistency are positive attributes in estimating and production.

Use Time-and-Materials Contracts

A business tool with less risk than a flat-price contract, a time-and-materials contract is a solid estimator's crutch. In a time-and-materials situation, the cost doesn't matter as long as it fits the customer's budget. The job simply gets marked up and paid for by the customer. You may do some early analysis to get a budget on the table for discussion, but you bill the work as you incur the costs. You should charge labor rates fully burdened and include a markup for overhead and profit. Materials handling charges may vary from 10% to 25% over costs. You would bill all materials pickup, special ordering, and supervision as hard costs against the job. A time-and-materials contract can decelerate the need to have all items specified early in the pricing. Their identification and pricing become critical when you order them. When you use a time-and-materials contract, you buy all the materials, supply all the labor, and bill the costs regularly.

A Project-Management Fee Contract

Another low-risk strategy is to charge a construction management fee on all work going into the job. This time-and-materials–type contract is slightly different. It provides for a global cost markup on all items going into the job, and generally the owner/customer pays all the invoices directly. A builder or remodeler would only touch the billable management fee. The customer may like the control, and the builder or remodeler may find the work more palatable, with less risk and lower accounts receivable and accounts payable.

Minimizing the workload is a combination of working fewer estimates, crisp systems, and delegating as much of the workload as possible. A team is available even to a one-person company. Look at the estimate preparation as a system of tasks, and look for the best person for each of these activities (Figure 7.3).

If you are a one-person team, you may think you are the only one involved, but in actuality you have a vibrant resource in all the vendors and tradespeople who will help construct the job. Taking advantage of their expertise should allow for increased efficiencies and some time-management opportunities. When you are buried under a pile of requests for estimates, the best initial strategy is to sort through them and see which ones are most likely to turn into deposit slips. Your pile of work just got smaller!

FIGURE 7.3 The Estimator's Team

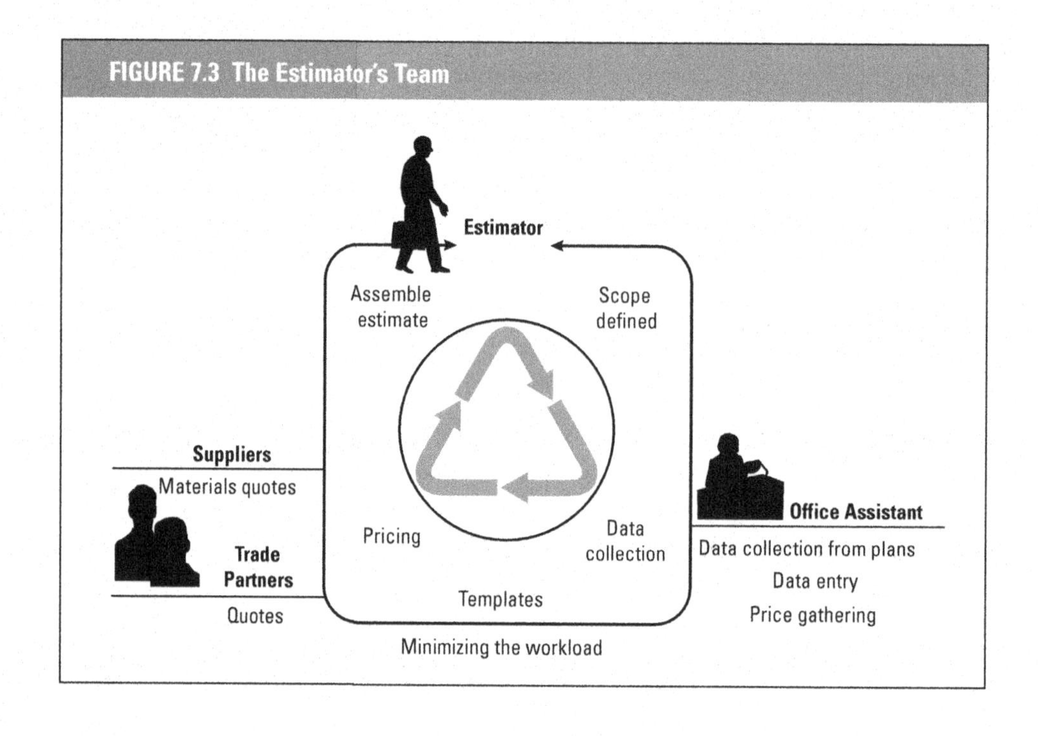

CHAPTER 8

Materials Costs and Tracking

Establishing a materials cost can only be done accurately in one way and that is to get a materials takeoff, including waste, and have it quoted by a reputable supplier who will hold the price for you. Assembling the materials list is time consuming and subject to clerical errors and errors of waste, layout differences, cutting excesses, and omission, to name a few pitfalls. The estimator sets out to attain a degree of accuracy that should be commensurate with how close the customer has come to a commitment to proceed to contract.

Often a simple guesstimate provides plenty of informational firepower to produce a qualified lead or dismiss the candidate as an underbudgeted consumer. This guesstimate or other shortcuts allow enough safety to keep moving the customer forward without doing full takeoffs for all the materials too early in the process. Based on fundamental time management, the details can come later as long as the estimate is tied to some estimator's "crutches," such as allowances, exclusions, or risk transfer. Plenty of contractors can't assemble an estimate until they have specified all the materials or completed a full takeoff. Although specifications or a takeoff are far more accurate, you want to avoid spending this overhead time before the prospect signs the contract. And you can easily defer it. By scaling the work dedicated to the prospect, you are actually defending your profits from erosion caused by lack of time management. If you track the annual hours you currently spend in estimating and compare that total to a scaled-back workload in preparing estimates, you could substantially reduce your estimating time. That reduction translates into less overhead and more company productivity. Given an overall industry closing ratio of 1 in 3 or less,

I met a multigenerational builder/remodeler some years back who was trying to establish his role in the family company. His grandfather had started the company and was still doing sales. The grandson, in his late twenties, related the story of going on his first sales call with his grandfather. They talked about the job and had a charming conversation with the potential client. The mentor and the neophyte estimator walked through the home then excused themselves to the front porch where grandpa lit a cigarette. After a few minutes he announced, "Okay, let's go give them the price." Startled, the novice asked what the price was and how the veteran had gotten it. He was told the number and assured that it was correct because he "knew what the job was worth." No database, no take-off sheets, no computer, but the right number, and the company flourished for decades.

the additional burden of a full takeoff has to be balanced against the likelihood of a sale and subsequent deposit slips. As the diagram in Figure 8.1 shows, you only need absolute clarity of materials selection and quantity when you are ordering as long as protections to profit are in place in the contract documents.

Astounding as the situation in the preceding sidebar may seem to today's truth seekers searching for the perfect number to present to a client, some "magic" exists in estimating that is the result of practice, time-tested experience, and human relations that gets the job sold and produced for the right numbers to produce a profit. This magical process does leave more risk in the equation, and it can potentially be devastating to the long-term health of the company. In the example cited in the preceding sidebar, the company was in a shambles several years after grandpa retired.

The numbers that fill the line items in a spreadsheet of cells, complete the database, or pencil in the lined legal pad can come from several sources. The guesstimate in the chart in Figure 8.2 stands out as providing the fastest and the least accurate method of estimating, whereas a full takeoff is the most accurate. The estimator's tools shown in the center of this diagram allow for the speed of shortcuts while they enhance accuracy over a guess.

FIGURE 8.1 Complexity of Materials Estimate

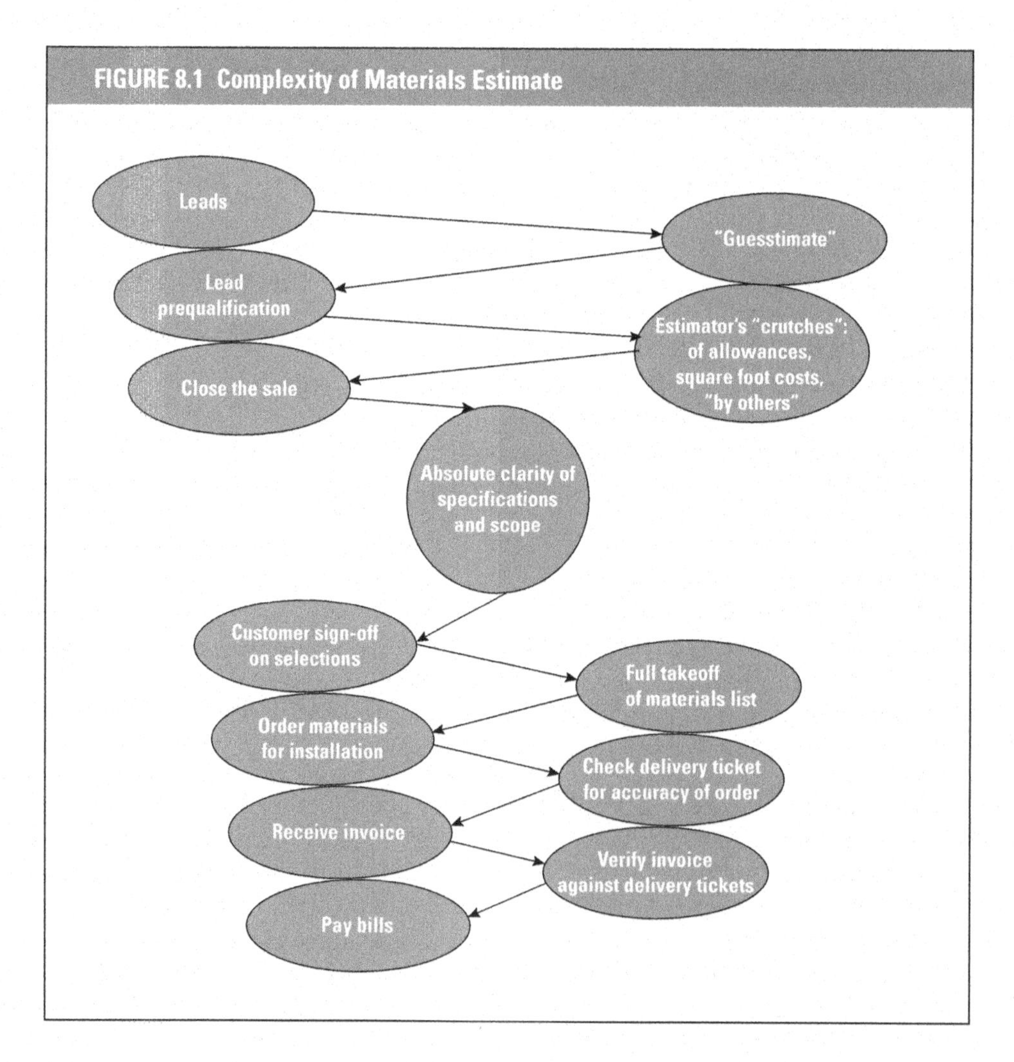

FIGURE 8.2 Where to Get the Numbers

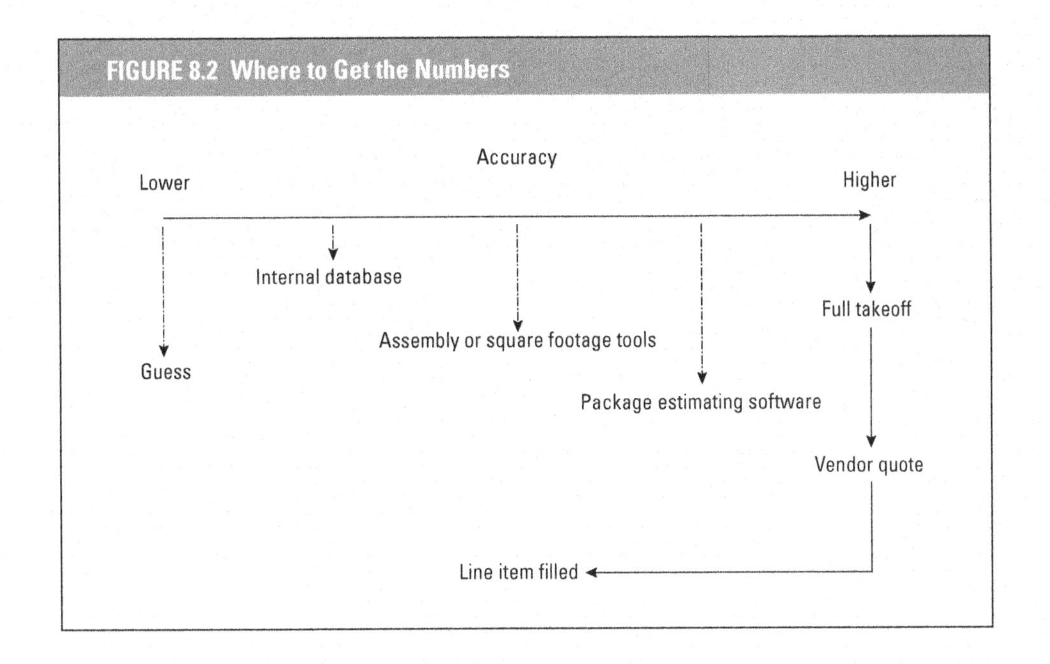

You need to build a system of data entry, monitoring, and feedback that can reproduce accurate results over time. The chronological look at estimating (as you would build the job) offers that framework. A breakout of materials for each line item is a simple way to accomplish a higher degree of accuracy than an overall price for materials. The greater the detail in the line items, the higher the overall accuracy of the estimate. I am reminded of a quote from a movie *The Patriot.* Actor Mel Gibson tells his volunteer soldier sons, when they are getting ready to ambush a patrol of Redcoats, to "Aim small; miss small." A small error in an extremely tight estimating subset will have little effect on overall job profitability.

When you are completing the estimate, this chronological system of data organization flows to the sale next and then into production. Production requires hard materials to work with so now detailing becomes more important. As you make purchases you need to tie them to the estimate or budget lines to measure the accuracy of the estimate and minimize discrepancies. The most efficient way to control materials costs against the estimate is with a purchase order system. This internal alignment of purchases matching the budgeted amount is extremely effective in controlling costs. You need to identify deviations at the commitment to purchase and not at the end the job or, worse yet, at the end of the tax year. Small-volume builders and remodelers could use a purchase order (PO) system because it offers the same controls that large-volume organizations need to succeed.

Simply stated, you number and assign a corresponding budget or estimate number to each purchase. Suppliers are accustomed to working with POs, and they can easily mark an account so that no purchases are transacted without a written PO number. The tracking alternative to a PO system is matching the bills to the estimate, but by this time, you are too late for cost control. The diagram in Figure 8.3 illustrates when you can implement con-

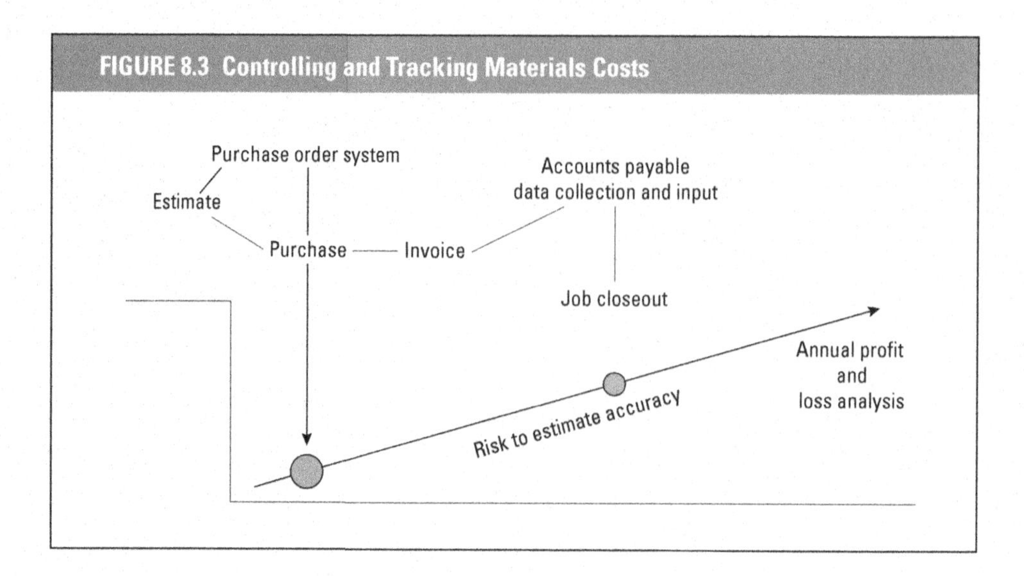

FIGURE 8.3 Controlling and Tracking Materials Costs

In our small business, we don't need to know exactly how many double joist hangers went into a custom job, but we do need to know if our lumber line item was sufficient.

trols for materials costs. The sooner you put the controls in place the less risk you have to profits.

Monitoring materials purchases for comparison to the estimate (estimate versus actual) uncovers deviations that you can correct in the next estimate. If you use POs, you generate each one from a line item in the estimate so you can readily monitor each one through invoice and payment. If you do not use POs, the tracking has to be done from the invoices, which means matching them to the estimate line. If you do this job through your accounting software, you have to set up the accounts payable with items that match those on the estimate that you want to track. As you saw in Chapter 4, Spreadsheet Estimates, in the sample spreadsheet in Figure 4.5, you can use subtotals for various assemblies. This level of tracking is more than sufficient to see if the framing lumber, for example, is over or under the estimate.

Each estimator must find his or her own comfort level with the amount of detail to enter and track. This entering and tracking could easily be an overwhelming task to calculate all materials and then monitor every item against the job.

Production work or time-and-materials work necessitates that you fully account for every item in the cycle. Every item you forget to bill in a time-and-materials contract is a cost to your company that is not reimbursable, so you get a double loss on the books: one for not billing the items and a second because you still have to pay for it. Custom building or remodeling may not mandate such an extensive overhead commitment because these activities are not likely to repeat on the next project, and general categories may be sufficient for tracking materials costs.

Tracking purchases can also uncover billable changes in the scope of work that you had limited with allowances. If you purchase tile at the home buyer's request at a price over the allowance numbers, the overage creates a billable recovery of costs, and thus you protect your profits. You should identify these changes at the selection of materials and get the customer to sign a change order for the overage rather than after the purchase. Without commenting in depth on the legal defensibility of billing cost overruns at the end of the job (a judge would require that the overruns be signed for before purchase), the additional step of getting client acknowledgement and approval for a cost overrun at selection of the item can save headaches later.

As previously discussed, the thinking estimator will look at materials numbers and attempt to protect profit from the ever-present risks. The trade contractors who roll materials into their prices take the materials cost risks from the builder or remodeler. You could regularly have many of your trade contractors include their materials in a fixed-price contract. Not only are their jobs simpler for the ability to control what they need and when they need it, but thereafter, you are not in the purchasing agent role for every item in the project. The trade contractors effectively become an extension of the builder's or remodeler's management team because the list for ordering and the time for shopping and delivery for the builder or remodeler has just gone to zero cost.

I think these shifts of risk and workload are healthy and an appropriate use of the full resources of all the trades working on the job. The builder or remodeler need not know how many ½-inch copper elbows go into the home or how many gutter brackets to get an estimate completed.

You need to defend against a number of risks to materials costs. These risks include inflation, materials shortages, distribution problems, delivery problems, storage issues, and credit terms. A market analyst may categorize the price risks trends as either technical or fundamental. A fundamental risk today includes regulatory changes and restrictions on supply that would raise prices. Conversely, opening new supply sources can drop prices. A technical trend would be to look at the rate of change compared to historical records, and if you see large differences, a correction back to historical norms is likely to occur.

The sample lumber chart in Figure 8.4 is followed by contract size (Figure 8.5) and a short market analysis (Figure 8.6) from the Web site http://futures.tradingcharts.com. Every estimator should log this site as a "favorite" so he or she has a sense about the stability of one of the largest-volume products. The components of the chart are price (low, high, and close by day) and trend lines over time. If you are buying quantities of lumber you would want to buy it with time and trend moving in your direction. This chart would be quite favorable to an estimate because prices are clearly dropping.

You need to know the current materials prices and how long the supplier can hold them. Suppliers are taking your price risk when they hold prices against a swirling market. You can manage price-volatility risk in a number of ways. One is to give the proposed contract an expiration date clause commensurate with how long the supplier will hold pricing. "This proposal is valid for 30 days" gives you a second look at pricing if the contract is not consummated within a safe window of price stability. You can also benchmark pricing to overall market rates. You can include a clause that states: "This proposal is based on Chicago Board of Trade lumber pricing as of _________, 20__. Any deviation from this base price will be adjusted as a change order on lumber materials." The National Association of Home Builders offers a standard escalation clause that can be used for a number of materials. Any escalation requires benchmarking to a starting price (Figure 8.7).

FIGURE 8.4 Sample Lumber Chart

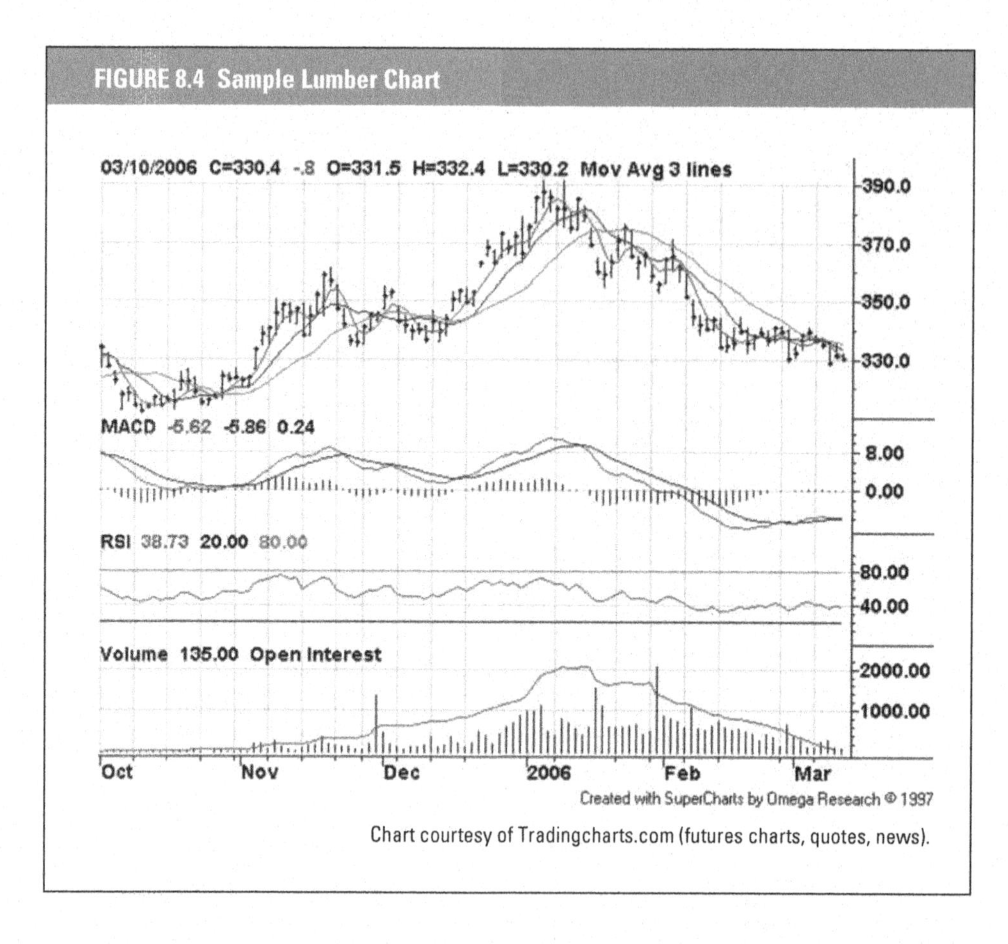

Chart courtesy of Tradingcharts.com (futures charts, quotes, news).

The National Association of Home Builders further limits risk to the profit line with a clause (Figure 8.8) relating to the substitution of specified materials. When the client changes his or her mind for a minor product, that change affects more than the product cost. The schedule, availability, possibly installation techniques, overhead time for documentation, and risk of col-

FIGURE 8.5 Contract Size

Contract Specifications: Lumber, Chicago Mercantile Exchange

Trading Unit:	110,000 board feet
Tick Size:	$.10 per 1,000 board feet ($11.00/contract)
Initial Margin:	$1,898
Maint Margin:	$1,265
Contract Months:	Jan, Mar, May, July, Sept, Nov
First Notice Day:	Business day after last trading date of contract.
Last Trading Day:	Business day prior to 16th day of the month.
Trading Hours:	9:00 a.m.–1:05 p.m. Chicago time, Mon-Fri. Trading in expiring contracts closes at 12:05 p.m. on the last trading day.
Daily Limit:	$10.00 per 1,000 board feet above or below the previous day's settlement price.

Courtesy of Tradingcharts.com (futures, charts, quotas, news).

FIGURE 8.6 Market Analysis

Conventional Interpretation: Price is below the moving average so the trend is down.

Additional Analysis: Market trend is down.

Mov Avg 3 lines Indicator: [No text provided for this subhead.]

Note: In evaluating the short term, plot1 represents the fast moving average, and plot2 is the slow moving average. For the longer term analysis, plot2 is the fast moving average and plot3 is the slow moving average. [These notes on the trend lines refer to the different divisors. A slow moving average might be divided by 200 days, and the short might be divided by 30 days.]

Conventional Interpretation—Short Term: The market is bearish because the fast moving average is below the slow moving average.

Additional Analysis—Short Term: The market is extremely bearish. Everything in this indicator is pointing to lower prices: the fast average is below the slow average; the fast average is on a downward slope from the previous bar; the slow average is on a downward slope from the previous bar; and price is below the fast average and the slow average. Warning. Market momentum slowed down on this bar. This [decline] is indicated by the fact that the difference between the two moving average lines is smaller on this bar than on the previous bar. . . . [A market rally may occur.]

Conventional Interpretation—Long Term: The market is bearish because the fast moving average is below the slow moving average.

Additional Analysis—Long Term: The market is extremely bearish. Everything in this indicator is pointing to lower prices: the fast average is below the slow average; the fast average is on a downward slope from the previous bar; the slow average is on a downward slope from the previous bar; and price is below the fast average and the slow average. Warning: Market momentum slowed down on this bar. This [decline] is indicated by the fact that the difference between the two moving average lines is smaller on this bar than on the previous bar. . . . [A market rally may occur.]

Conventional Interpretation: The Bollinger Bands are indicating an oversold condition. An oversold reading occurs when the close is nearer to the bottom band than the top band.

Additional Analysis: The market appears oversold, but [it] may continue to become more oversold before reversing. Look for some price strength before taking any bullish positions based on this indicator.

Courtesy of Tradingcharts.com (futures, charts, quotas, news).

lection also are affected—to name a few. This clause can save the day against any number of profit detractors. Estimated costs must be utilized and defended. The scope is as important as the price in an estimate, and this clause helps solidify the scope.

The supply chain for construction materials is typical of any other market in which supply capacity and demand affect current pricing. Natural

FIGURE 8.7 Escalation Clause for Specified Building Materials

The contract price for this residential construction project has been calculated based on the current prices for the component building materials. However, the market for the building materials that are hereafter specified is considered to be volatile, and sudden price increases could occur. The [Builder or Remodeler] agrees to use his best efforts to obtain the lowest possible prices from available building material suppliers, but should there be an increase in the prices of these specified materials that are purchased after execution of contract for use in this residential construction project, the Owner agrees to pay that cost increase to the [Builder or Remodeler]. Any claim by the [Builder or Remodeler] for payment of a cost increase, as provided above, shall require written notice delivered by the [Builder or Remodeler] to the Owner stating the increased cost, the building material or materials in question, and the source of supply, supported by invoices or bills of sale.

	Specified Building Material	Current Price Per Unit of Measurement	Date	Supplier
1.	______	______	______	______
2.	______	______	______	______
3.	______	______	______	______
4.	______	______	______	______
5.	______	______	______	______
6.	______	______	______	______

Special Circumstances—Right of Termination

Should there be a rise in the cost of any specified building material or materials, exclusive of any other price changes, that would cause the total contract price to increase by more than _____(%), the [Builder or Remodeler] shall, before making any additional purchases of specified material or materials, provide to the Owner a written statement expressing the percentage increase of the contract price, the building material or materials in question, and the dollar amount of the price increase to be incurred. The Owner may then, at his or her option, terminate the contract by providing within _____ business days both written notice of termination to the [Builder or Remodeler], and payment to the [Builder or Remodeler] for all costs expended in performance of the contract to the date of termination, plus payment of a prorated percentage of the Builder's profits based on the percent of completion. Should both notice of termination and full payment not be forthcoming within _____ business days, as provided herein, the [Builder or Remodeler] shall have the option to terminate the contract, or to proceed with the contract and purchase the specified building materials at the increased price. If termination is elected, the [Builder or Remodeler] shall provide to the Owner a written notice of termination, and the Owner shall be required to pay the Builder for all his costs expended in performance of the contract to the date of termination, plus payment of a prorated percentage of the [Builder's or Remodeler's] profits based on the percent of completion. If the [Builder or Remodeler] elects to proceed on the contract, he may then purchase the specified material or materials at the increased price, and the Owner shall be required to pay the increased cost incurred.

Caution. The sample language provided in this clause is intended for general informational purposes only, and it may not be appropriate for some agreements. Care should be taken in the drafting of any contractual clause, and you should consult an attorney concerning both applicable law, and the phrasing of particular contract provisions.

disasters, environmental regulations, code requirements, production constraints, and to a lesser extent, foreign competition can affect the current market. This ebb and flow of price instability is fodder for entire books of study. However, an estimator should be familiar with the overall direction of materials markets. Market pricing and trend information was once the sole domain of brokerage specialists, but the Internet now provides both up-to-the-minute access and historical charting information on price trends for free

FIGURE 8.8 Substitution of Specified Materials

Whenever the contract provides for the use of specified building materials or fixtures, the [Builder or Remodeler] shall be required to use the materials and fixtures as specified, except as follows:

1. In the event that a specified building material or fixture cannot be obtained in a timely fashion because of nonavailability or a shortage in supply, cannot be obtained at a reasonable cost, or cannot be obtained for any other reason that is not the fault of the [Builder or Remodeler], the [Builder or Remodeler] shall be permitted without additional approval to use substitute materials or fixtures of equal or better quality and performance as rated by ____________ [specify the rating service]. "Timely" shall mean that no work stoppage is incurred, or delays are experienced in the construction schedule. Reasonable cost shall mean not more than ___% additional cost of the material or fixture above its cost at the time the contract is signed.
2. The Owner shall be responsible to the [Builder or Remodeler] for any additional cost (over and above the supplier's price for the specified materials), and any additional expense that is incurred by the [Builder or Remodeler] in obtaining the substitute materials. Provided, that in instances where the specified material cannot be obtained at a reasonable cost, the Owner shall not be responsible for any additional cost or expense that exceeds the then current price of the specified materials or fixtures.
3. Any delay incurred as the result of a failure to obtain a specified building material or fixture that is through no fault of the [Builder or Remodeler], including any time that is necessary to obtain a substitute material, shall result in an extension of the time of performance under the contract for a period equal to the time of delay. In no case will such an excused delay be considered a breach of contract.

with a few clicks to the source. Controlling market risk in the supply channel is called *hedging*. Entire teams of buyers at the large suppliers have as full-time jobs reducing market risk from materials price swings. Interest-rate hedging was used more in prior decades when rates seemed far more volatile. Stability can be an elusive and short-term attribute in business when interest rates ratchet upward. Market elasticity can absorb mild price fluctuations, but large unprotected swings can be fatal.

Seldom will a builder or remodeler get too much material in error, but our history of matching delivery tickets to a field count certainly shows that we occasionally get a short order. Minimizing these shortages protects profits and deflects the expense of a large amount of on-site handling of replacement materials.

A materials cost deviation can also occur at the jobsite when materials disappear because of theft or damage because of a lack of protection. Materials can easily be looked at as money on the site and should be handled accordingly. You need to check every delivery for accuracy and report deviations.

As discussed in Chapter 4, "Spreadsheet Estimates," when you can roll materials costs into a trade contractor's installed price, you stabilize the materials costs against his or her contract. You have many ways to keep the risk of materials purchases at bay. Another strategy may be to get longer-term commitments from supply houses in exchange for regular purchase activity.

Inflation has been tame the past few years, but it could be back in the builders' or remodelers' fields of vision, and you must avoid its impact or

Our personal bias is to use the suppliers as vendor partners and not merely as a conduit for the lowest cost materials.

transfer it. A conventional way to stay sharp on materials pricing is to get multiple vendors to bid on the materials with a separate price for each job.

A custom builder or remodeler can make great use of the services of the supply chain for delivery, return policies, education, takeoff assistance—to name a few attributes, and these no-cost "value adds" are quite attractive when put to good use. A strong supplier relationship can help protect against damage to profits from a price increase when a particular job is at risk. Clearly, market prices of materials change. Therefore, a current estimate will reflect current prices and reduce the risk that you will have to pay for a price escalation with profits.

CHAPTER 9

Estimating and Tracking Production

For assessing risk to profit, estimating production is likely to be the toughest segment in the entire construction process. The variables of weather, materials availability, labor skills, jobsite conditions, and more are lurking just around a dark corner and are only satiated with a bite out of profits. The simple answer to quantifying costs of production is to contract for a fixed amount of work for a fixed price. Work completed by employees does not have the stability of price that a subcontract may hold.

A subcontract shifts the responsibility to the trade contractor to perform at a certain dollar amount. Even if you are engaging a trade contractor for a portion of the work, you still need to cover some reasonable vagaries to a contract and scope of work to keep risk at bay. An assessment of whether to do the work in-house or contract it out requires an evaluation of production costs versus a fixed price. The fixed price subcontract will include overhead and profit for the firm's owner.

Given an equal skill level, completing the work in-house should be to the builder's or remodeler's cost advantage. The qualifier, an equal skill level, is often tough to assess. The tradespeople doing repetitive work day after day are likely to work not only faster but also simply better than in-house labor.

Solid production supervision can equalize some of the quality concerns between in-house and contracted labor. A sound way to assess production values early in your estimating career is to do a series of plan reviews and site inspections with qualified trade professionals. Such reviews and inspections will help you to understand what they are looking for at the jobsite. A qualified trade professional can also identify any unforeseen conditions and alert the estimator to pitfalls on the jobsite or within the scope of work specifications.

You could use any of a number of productivity databases available for review through trade publishing houses. These databases show an approximate time to calculate and use in the estimate to complete certain work. These databases could include how many square feet of dropped ceiling a worker can install in a day or the number of blocks that a two-man team can lay in a day. If you multiply these units of productivity by the rate of pay, you can determine what your costs should be. Understanding site conditions and work specifications is critical to getting the productivity numbers close to accurate.

Increasing accuracy in estimating production results from consistent and accurate techniques in the field. When estimating a concrete pour at 4 inches

I recently had a small addition to complete, and the trim for four rooms took seven months. Each piece was hand machined out of oak and installed. We constructed multiple profiles to add interest. Because we recognized the difficulties of defining this scope early, we utilized one of the estimator's crutches and took the job on a time-and-materials basis, which shifts the risk to the clients. They got a good day's work and an end product of their choosing, and we avoided getting clobbered with runaway production costs that were not anticipated.

thick, if the gravel preparation is set at 5 inches below the finished floor line, the volume of the pour is in error by 25%. One inch may seem insignificant in a 30 × 40–foot basement floor, but the resultant overrun would leave you with nearly 4 yards of additional material at roughly $100 per yard plus the additional labor costs for handling the installation. Thus, a 1-inch preparation tolerance could approach an error of $1,000 for lack of field controls to match the estimator's expectations with field production.

Lots of communication and a full understanding of what you sold your client are irreplaceable. If you sell a vanilla specification, the freedom in the field to do outstanding custom work will surely create cost overruns.

The time you include in the schedule when you are estimating work flow should result from a consensus of your employees and/or trade contractors and construction management. An unrealistic schedule helps no one. Once you obtain consensus on the schedule, tracking and evaluation need to take place (Figure 9.1).

FIGURE 9.1 Tracking Production

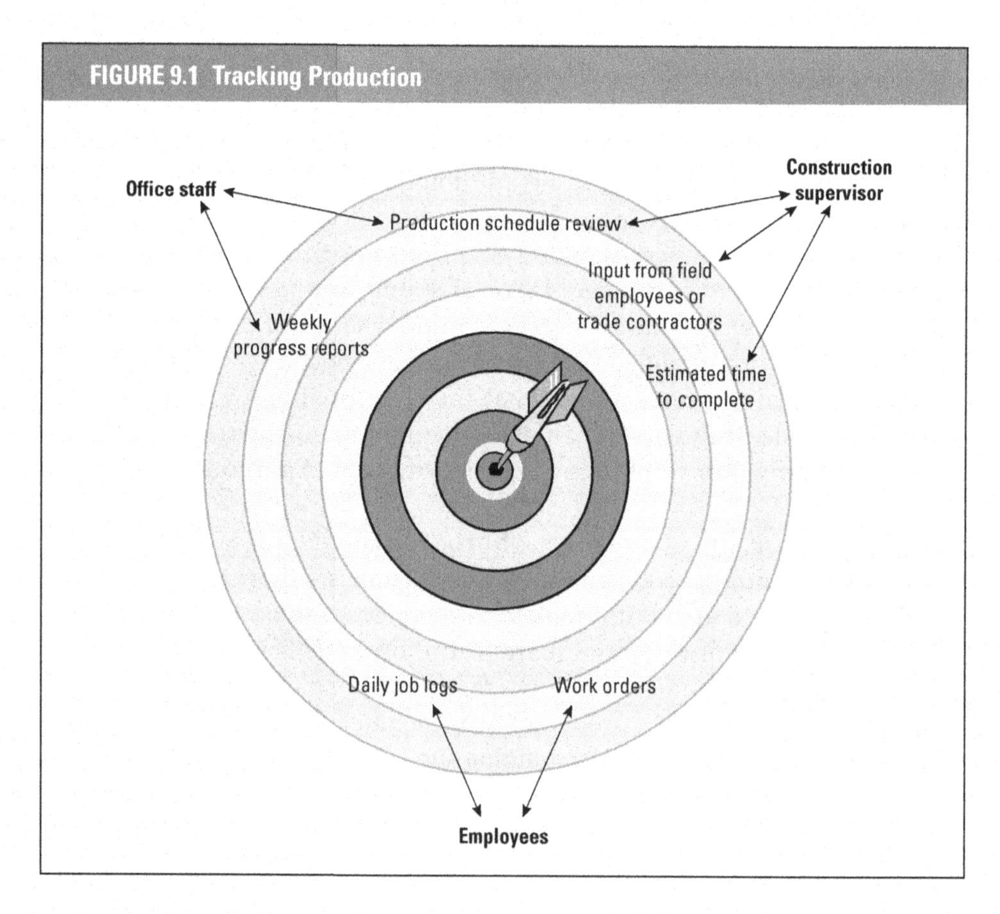

The estimated time considerations have been a factor of the costs you put into the estimate. The time multiplied by rate equals the cost. As you construct these production costs, you can use the same data to build the construction schedule. Herein lies the beauty of estimating tasks chronologically. The same system creates both the estimate and the timeline.

Communicating Between the Jobsite and Office

To turn the responsibilities into tasks, write work orders that describe what is to be completed. You may include an expected duration on the work orders, and certainly, they should contain a detailed scope of work. Lack of clarity on the work orders foretells of conflicts between what was estimated and what actually gets done. Field labor can easily be distracted to other tasks, and while they are productive, they may not be moving on the anticipated timeline. In critical-path scheduling of activities, this situation would delay completion. (More on construction scheduling is available at www. BuilderBooks.com.) If a work order provides the paper trail from the estimate to the field, the returning document is a daily job log (Figure 9.2).

The job log provides communication back to the office staff. The name of the employee and the jobsite name and address are indicated at the top of the log. For each hour of the day the employee enters a cost code from the list on the right that relates to a series of activities with the same coding on the estimate spreadsheet. These cost codes tie back to the spreadsheet for accountability when the estimator is completing a job-profitability report. This simple report uses the job log information (input for payroll) to compare estimate versus actual cost for each line item that is coded.

We also print out this activity report for our workers' compensation insurance audit. Not by accident, the labor divisions on the daily job log are the same as the New Jersey Division of Workers' Compensation. The rates for insurance coverage vary by the activity each employee is performing. When we can provide a report that details where every employee has been and what they have been doing for each hour of the year, we are able to divide our labor into the workers' compensation cost codes so that our insurance rate is based on worker activity. The workers' compensation insurance rate for carpentry is just above $10 for each $100 of compensation, whereas supervision is $3 per $100. When we can prove an employee's roll is supervisory, the rate drops significantly.

For our purposes using the subsets of estimated numbers is adequate. You will find these subsets in Chapter 4 as subtotaled sections of the full estimate spreadsheet (Figure 4.5). You can compare the estimated cost of each activity to its actual cost. You also can filter this report by employee to see how much time each is spending in various activities.

Reporting by work item is another way to track productivity. Similar to tracking by employee, this self-reporting method focuses on the line item. In the work-item method, the bookkeeper codes the payroll hours as items, and they can then be compared to projections in the estimate. Whether sorting by job, activity, or employee, a much better financial picture of the work is visible.

FIGURE 9.2 Daily Job Log

Your Building Company, LLC

Daily Job Log

Employee Name: ____________________ Day and Date: ____________________

Weather: ____________________

Job Name: ____________________ Temperature: ____________________

Job Location: ____________________

Start of Work:	**Work Activity**	
	8	02 Demolition
		03 Site Work
	9	03.10 Site-Work Demo
		04 Excavation
	10	05 Concrete
		06 Masonry
	11	07 Framing
		08 Roofing/Flashing
	12	09 Exterior Trim
		10 Doors/Windows
	1	11 Siding
		12 Plumbing
	2	13 HVAC
		14 Electric/Lighting
	3	15 Insulation
		16 Drywall
	4	17 Ceiling/Coverings
Completion of Work		18 Millwork/Trim
		19 Cabinets/Vanities
Others on Job:		20 Floor Cover
Name:		21 Paint
Work Completed:		22 Cleanup
Hours:		23 Landscaping
		25 Supervision/Pick Up Materials
Name:		
Work Completed:		
Hours:		

Phone Calls:		Nature of Call:	
Phone Calls:		Nature of Call:	
Phone Calls:		Nature of Call:	
Materials Ordered:		Quantity:	

Remarks:

No injuries unless otherwise noted here:

If you figured 22 hours for siding work in the estimate and the results entered for payroll for the work item actually totaled 24 hours, part of the job variance in labor was this 2 hours. Materials could also be lumped in at the item level or done separately. Coding activities could be daunting, but using a daily job log filled out by each employee makes coding simple and efficient. The item numbers in the accounting and payroll software are the numbers the bookkeeper takes from the daily job logs (Figure 9.3).

Changes in Scope

The construction manager needs to have in hand clear communications regarding what work items are to be done as part of the base contract and what activity may trigger a change order (Figure 9.4). Each employee should keep in his or her possession a simple blank change order so that each employee can record any additional work and trigger a billable event. The change order language allows for work to continue without interruption and defends the profit line against erosion from paying for additional work that exceeds the estimated scope. You should obtain a signature from a responsible owner at the earliest convenience—ideally before any of that work is started—to cement the obligation of the owner to pay for the additional work required.

We waive the minimum fee noted on the change order form when we use it for selections. We create a selections schedule from the estimate form line items that offer a customer choice.

You should give the home buyer or owner a selections schedule at the start of the job and provide updates as progress continues in an attempt to stay current and accurate. You also can enter selection decisions on the change order form with a signature to document choices.

You should give designated employees authority for contact with the client for decision making, and the employees should document all changes in writing and file them in a single job-specific file.

Production Benchmarking

When you are using production data in the estimate to figure costs, you need to check how the work is progressing. To tell exactly how work is progressing statistically is nearly impossible. When is a roof half done? When half the shingles are on, or does delivery and set-up contribute heavily to upfront costs incurred? In residential work, these interesting benchmarks affect progress payments and cash flow for large jobs.

A construction supervisor or builder or remodeler should explain expectations for the field crew before the start of work. The explanation could be as simple as, "I expect this work will be done tomorrow." If the field crew buys into the objective, and the work continues past the projection, the crew needs to account for the added time. You can track productivity using defined milestones. How much work do you expect can be done within a time frame and produced at a specific level of performance? Tracking can be as simple as a nod and a "thank you," or it can be as thorough as entering status updates on a schedule.

(Continued on page 92)

FIGURE 9.3 Payroll Data Entry

Item		Description
01	Plan-Perm	Plans & Permits
02	Demolition	Demolition
02s	Demolition	Demolition by Trade Contractor
03	Site Work	Site Work
03.10	Demo	Demo
03s	Site Work	Site Work by Trade Contractor
04	Excavation	Excavation
04s	Excavation	Excavation by Trade Contractor
05	Concrete	Concrete
05s	Concrete	Concrete by Trade Contractor
06	Masonry	Masonry
06s	Masonry	Masonry by Trade Contractor
07	Framing	Framing
07s	Framing	Framing by Trade Contractor
08	Roof-Flash	Roofing, Flashing
08s	Roofing-Flashing	Roofing by Trade Contractor
09	Ext Trim	Exterior Trim & Decks
09s	Exterior Trim	Exterior Trim by Trade Contractor
10	Doors/Windows/Trim	Doors/Windows/Trim
10s	Windows/Doors	Windows/Doors by Trade Contractor
11	Siding	Siding
11s	Siding	Siding by Trade Contractor
12	Plumbing	Plumbing
12s	Plumbing	Plumbing by Trade Contractor
13	HVAC	Heating & Cooling
13s	HVAC	HVAC by Trade Contractor
14	Electric	Electric
14s	Electric-Light	Electric by Trade Contractor
15	Insulation	Insulation
15s	Insulation	Insulation by Trade Contractor
16	Drywall	Drywall
16s	Drywall	Drywall by Trade Contractor
17	Ceiling-Cover	Ceilings & Coverings
17s	Ceilings/Coverings	Ceilings/Coverings by Trade Contractor
18	Millwork-Trim	Millwork & Trim
18s	Millwork/Trim	Millwork/Trim by Trade Contractor
19	Cabinet-Vanity	Cabinets & Vanities
19s	Cabinets/Vanities	Cabinets/Vanities by Trade Contractor
20	Floor Cover	Floor Coverings
20s	Floor Coverings	Floor Coverings by Trade Contractor
21	Paint	Painting
21s	Painting	Painting by Trade Contractor
22	Cleanup	Cleanup & Restoration
22s	Cleanup	Cleanup by Trade Contractor
23	Landscape-Paving	Landscape & Paving
23s	Landscaping	Landscaping by Trade Contractor
24	Miscellaneous Work	
24s	Miscellaneous Work	Misc. Work by Trade Contractor
25	Supervision	Supervision

FIGURE 9.4 Change Order Form

Your Company Letterhead

Change Order Form

Name: ______________________ Work Begun: __________

Street Address: ______________________ Work Completed: __________

City and State: ______________________ Zip Code: __________

Additional Work Description:

Additional Materials: ______________________

____

____

Additional Time: required for work that will extend contract completion: __________days

Total Price Prior to this Change $ __________
Additional Cost: $ __________
Total Revised Price __________
Revised Schedule of Payments:

____

____

____

____

You are hereby authorized to make the change(s) listed above in your work on the original Agreement, and we understand that this additional work will be executed under the terms of and the conditions embodied in our contract. This Change Order is subject to final approval and acceptance by [insert your company name] . All guarantees and warranties shall be consistent with the terms in the contract. Payments for any and all additional Change Order(s) are due and payable prior to commencement of such extra work and upon signing of this Change Order Form. All Change Orders have a minimum fee of $60.00 for overhead and processing. In the absence of a fixed price, work will be billed on a time-and-materials basis at rates defined in the base contract.

The estimated completion date provided for in paragraph ____ of the contract is now ______(date)______. All other terms and conditions of the contract referred to above remain unchanged.

______________________ ______________
Owner Date

______________________ ______________
Production Supervisor Date

Following the interim tasks between start and completion dates of a job underway can help you better define when completion and occupancy can take place. This tracking could be in units of measure, units installed, per piece, or compared to the trades currently on the site. Keep in mind the fact that you used full retail rates in the estimate. Therefore, small delays may dent the line item profit, but they will not touch the planned job profits. Back to Figure 9.1, the cyclical nature of estimating, production, monitoring, and review positions the estimating task for the next job with a clearer vision of what the costs will be.

CHAPTER 10

Financial Analysis: Estimating the Cash Flow

The purpose of estimating (as the theme runs through this book) is to minimize risk to profits. The risk involved in some tasks are easy to minimize—for example, the numbers within the estimate—because the builder or remodeler has nearly full control of the costs, or he or she also can deflect unexpected costs with precise language.

Customer risks are much harder to control. You need to avoid them through careful analysis and firm application of sound business policies. For example, how long do you allow an overdue bill from you to go unpaid? How do you anticipate and control damaging indecision? Should you get a credit check on the customer to ensure you are signing the contract with an honorable party whose track record in payments is solid?

You also have regulatory risks; market risks; and safety, production, and inflation risks. The risks to the financial aspects of the building and remodeling industries include not only whether you will receive enough money from the job to be profitable, but also the risks to cash flow. Being profitable on paper and having no money makes no sense. You need cash to keep the job running efficiently.

"Cash is the gas . . . that makes a building company run."

—Alan Hanbury, CGR, CAPS, House of Hanbury Builders, Inc., Newington, Connecticut[4]

Projecting the Cash Flow on the Job

Because the estimate template flows in chronological order, you can easily align the costs of the job with the proposed client payment schedule. If the client is working with a bank, the bank would call this payment schedule a *draw schedule* because the loan amount is "drawn" down and paid through the client to the builder or remodeler.

No matter what the client's source of funds, the amount paid to you as the builder or remodeler must align with or exceed the costs incurred to get to the next payment in the schedule. More specifically, you should run the job with the client's money through the entire job. Even if the supply houses give excellent credit and liberal terms, their money should not be going into your clients' homes. Neither should the trade contractor's funds, your overhead, or any materials go into the job before progress payments are received. The organized sequence of estimating tasks shows when the bills will be coming in, so adequate cash needs to be billed to the customer to make these payments align with anticipated expenses.

Working with the owner's bank and having some input on how the draw schedule is structured can be critically important for defending your profits. The items on the estimating spreadsheets show what funds you need to get to a given point. The estimate shows some deviations from the estimate to actual costs, but these deviations are not problematic here because the spreadsheet in Figure 10.1 illustrates cash flow.

You should bill against the current work the amount of money necessary to get to the next draw. This practice of estimating cash flow and billing proportionate to expenses ensures enough money is flowing in to prevent financial disruptions of the jobsite. Well-funded jobs always move faster than underfunded ones because you can pay the trade contractors sooner and they have an added incentive to get their work done. Nearly everyone appreciates quick payment on completed work. In the table in Figure 10.1, the estimated costs and payments received illustrate that the initial deposit needs to be sufficient to get to the first payment. Subsequent draws also need to be sufficient to get to the next payment. The table shows six payments over an 18-week job.

FIGURE 10.1 Sample of Estimate Versus Actual Expenses Report

Any Job Contracting, Inc.

Cash Flow Analysis	Estimated Costs	Expenses Incurred	Payments	Cash Flow
Week # 1	$ 2,000.00	$ 1,500.00	$10,000.00	$ 8,500.00
2	$ 5,000.00	$ 4,500.00		$ 4,000.00
3	$ 3,200.00	$ 3,800.00		$ 200.00
4	$ 5,000.00	$ 4,500.00	$16,000.00	$11,700.00
5	$ 4,000.00	$ 4,500.00		$ 7,200.00
6	$ 2,000.00	$ 1,600.00		$ 5,600.00
7	$ 6,000.00	$ 4,000.00		$ 1,600.00
8	$ 4,000.00	$ 5,000.00	$16,000.00	$12,600.00
9	$ 1,200.00	$ 1,200.00		$11,400.00
10	$ 2,000.00	$ 3,000.00		$ 8,400.00
11	$ 4,000.00	$ 4,000.00		$ 4,400.00
12	$ 4,000.00	$ 3,700.00	$16,000.00	$16,700.00
13	$ 2,000.00	$ 3,400.00		$13,300.00
14	$ 4,000.00	$ 5,300.00		$ 8,000.00
15	$ 1,500.00	$ 1,400.00	$15,000.00	$21,600.00
16	$ 5,000.00	$ 3,500.00		$18,100.00
17	$ 2,000.00	$ 700.00		$17,400.00
18	$ 1,000.00	$ 1,300.00	$ 5,237.75	$21,337.75
	$56,900.00	$56,900.00	$78,237.75	
Overhead	$ 5,690.00			
Subtotal	$62,590.00			
Profit	$15,647.50			
Sale Price	$78,237.50			

The chart in Figure 10.2 shows that the cash flow stays positive as the job progresses. The heaviest line curving upward in the chart shows the remodeler's or builder's position relative to cash produced by the job. That position gets better as the work gets completed. The curve appears to be shaped like a smile.

The finer curved line represents the client's payments. They begin modestly, get more substantial as the job is built, and trail down to a smaller final payment. The client-payment curve looks like a frown.

Note that in the chart the cash-flow balance does not dip below $0.00 so no outside cash goes into the construction of the job. Another observation of the chart shows that the final payment ($5,237) is less than the positive cash-flow position from the prior payment ($17,400). In the unlikely event that someone gets grumpy at the end of the job and doesn't make a final payment, that situation may hurt the profit but not the builder's or remodeler's business. This safe position defends your company.

Estimating cash flow is critical to a well-run company because shortfalls slow the job, and accelerated payments can speed completion and enhance profits. In this regard, you can determine to make more money with a well-

FIGURE 10.2 Simple Cash Flow Model

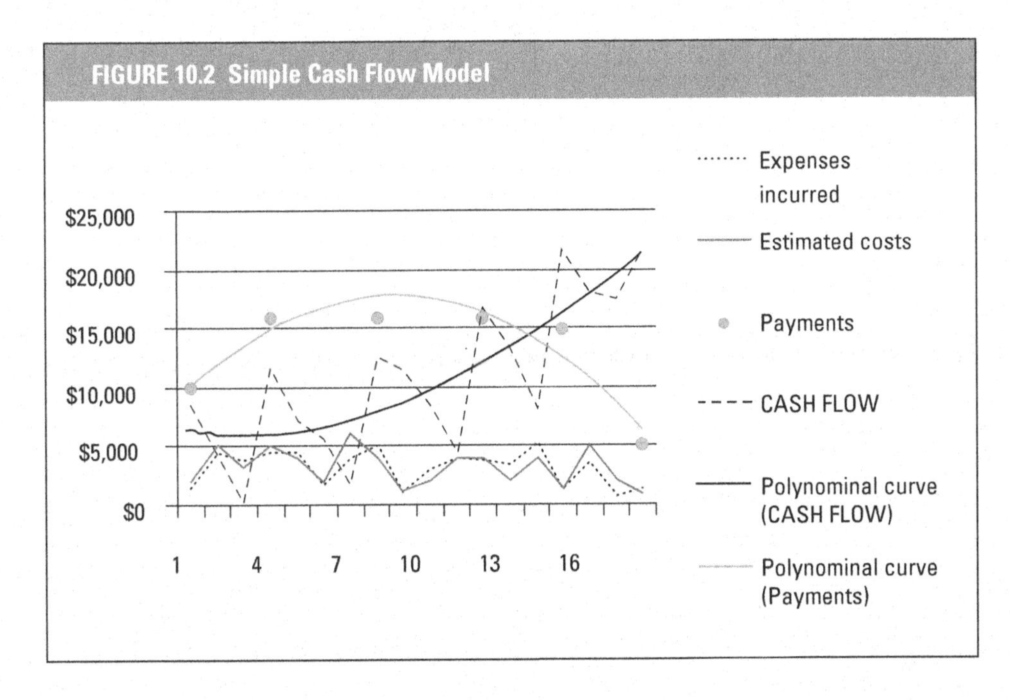

funded job. Defensive estimating is the system that provides the cash to fuel profits.

In observing the line clusters near the bottom of the chart, the expenses incurred and estimated costs are plotted in a generally narrow band showing no wild deviations of estimated and actual costs. This general alignment keeps the cash flow projections accurate. A major error for any line item can turn the cash flow upside down because you would have to pay for mistakes. When you are defensively estimating, you shunt the risk of errors in line items to others or offset that risk with a change order that generates compensating payments and keeps cash flow intact.

A major hit to cash flow comes from a retainage clause in a building or remodeling contract. In the draw schedule described above, little to no builder's or remodeler's money should be in a custom home or a remodeling job. A retainage clause will withhold some percentage of every draw through the contract life. If the net profit planned is 10% and the contract calls for a 10% retainage to be withheld, you must make up the shortfall elsewhere because you will need the hard cash to pay your bills. This accumulated retainage is not released until the end of the job, and occasionally release occurs after a warranty period. You should suppress the temptation to sign anything other than your own defensible contract. Many of the contracts in circulation have plenty of clauses that affect compliance, influence pricing, and define notices that are not builder or remodeler friendly. The terms and conditions in your contracts need to clearly defend your interests.

Reports

When you compile the cash flow records from multiple jobs, you get a great look at cash flow for your company. You will encounter inevitable troughs in funds that you can manage with credit and vendor trade discounts or simply by accelerating payments when necessary. You can report your cash positions with a statement of cash flow. To keep your contract cash-flow positive, tighten the turnaround time for receivables and collect promptly.

You need an estimate-versus-actual report to keep your database current. If you find deviations, review them to see why and who was responsible, then update your database or formulas. For example, if you are constantly doing highly profitable concrete work, you may think of taking this niche to the market place so you do more of it. Conversely, underpricing a line item could be as simple as an estimating error, you could have a materials issue, or your labor may not prove as productive as anticipated.

Finding these deviations is a management task, and realigning them will take insight and willpower. Identifying these *variance action items* in your budget could easily consume a full-time position, and it is likely to be a major concurrent responsibility of jobsite supervision. If the same person is both the estimator and the field supervisor, corrections need to flow back through your systems to realign output with expectations. What is the field crew doing differently from what you estimated? This basic question is the crux of running profitable jobs. Sharing expectations of how long you think a job will take to complete with the field team provides appropriate help for them to gauge progress.

When you are job costing and tracking expenses to compare to the estimate, you need to post these costs to the ledger for the specific job for which you incurred them. Seldom does anyone use separate job ledgers because accounting software can run a report in seconds as long as it can trace the data entry of receipts and expenses to the job for which you purchased the items and/or used them. The point is to set up the accounting to match the estimating. So, if the estimating is exactly as you would like it, alter the accounts or create subaccounts that mirror your systems.

You can import spreadsheet estimates into your accounting software, a handy way to eliminate manually transferring the data. Using subcategories that combine a number of small line items in the estimate for tracking shortens the bulk of the estimate data that you need to transfer. For example, the line item for cleanup could include debris removal and the cost of renting the dumpster. These groups of items can be more than sufficient to track expenses against estimated costs. The posting of each invoice should include the job name and the activity. These two data points are the basis for, and are critical to, any effective tracking.

To help both the estimator and management, create a repetitive process of formal reporting requirements and mandatory review techniques. Collect the same reports to check on the same estimating techniques each time an estimate goes through the production cycle. You may need to update or correct your estimate formulas, review your site inspection techniques, begin having another person double-check your work, or update your price sheets. Because no magic message tells you how frequently you need to update spreadsheets, you should make this review continual and ongoing. Simply getting to a profitable job does not mean the processes are efficient. A series of profitable jobs would imply that the processes are working.

Discussing the financial exposure of a job with employees, trade contractors, and suppliers is appropriate. If you are open and dialogue about the risks, the specifications, and details, everyone can be clear about the scope of the work and budget accordingly. What is the worst incident that can happen? Nonpayment, divorce of the customers, change of plans, health-related issues arising out of the construction, and a warranty claim are each a candidate for a nightmare script. Any of these events could possibly happen over time, and the most unexpected event may become the one most likely to occur. You need to create and implement a defensive strategy that minimizes risk from any of these occurrences.

What are the systems and forms necessary to minimize risk? The right consumer contract, the right purchase order wording, and a clear trade contractor agreement are a few of the dozens of pieces that you can create so that your business breaches no bottomless pits of risk. You need to hone your internal company systems to a defensive posture and enhance your personal skills to help you defend your profits. A memo in the job file with a date stamp after a conversation can go a long way if you ever have to defend

Educating employees that receipts can only be entered if they have the job cost codes and the clients' names on the receipts should help decrease omissions. Data entry is made easier and coding of items is far more accurate when the field employee is engaged in recordkeeping.

your decision to act in a certain way. Taken to the next level, a letter confirming a conversation is a great way to benchmark a decision. It may read:

> Dear Building Inspector:
>
> This letter confirms your oral approval of today's footing inspection. . . .

As we have rolled out defensive estimating for protecting your profits, a builder or remodeler can also begin thinking about the defensive office.

Faxed to the inspector's office and put in the job file, this confirmation assures that documentation supports the activity of the day. Future "he said, she said" confrontations are immediately laid to rest with a defending note in the job records. You can use the same technique with verbal quotes or consumer selections.

Personal skills that defend profit may include how you interact with your clients. To defend yourself and your company against elevated expectations, be careful not to overpromise at the sales level something your production team cannot produce on the jobsite. Defensive traits, defensive business systems, and yes, defensive estimating should secure profits that fund personal goals. You have a formula for fulfillment at hand.

CHAPTER 11

Defending the Profit Line in Your Building Estimate

Some components in the estimating process are inherently more risk laden than others. The graphic in Figure 11.1 is the diagram in Figure 1.2, "Workflow in Estimating," with an overlay of potential problems that could destroy profits. Note that the figure shows no risk to profits in the process of prequalifying customers. The only danger at this point is failing to eliminate unqualified leads early enough. As a result, you may find yourself doing too many estimates and courting unqualified clients far too long before detaching from them. This situation surely impacts overhead costs because your time is valuable, and you could be doing something that would add to the profit line item for your business. The builder's "lead to close" time is long and expensive to maintain. You need to quickly identify and cultivate promising candidates and cull the unlikely leads.

Data collection is the first activity in which estimating can start to go wrong. If you fail to observe a critical piece of information or misread the field conditions, these oversights may begin the long road to an unprofitable job. A builder needs to be observant of existing conditions that can affect the scope of the work. These omissions create a deficient contract, a work in progress requiring line item costs that you may not have included, and attempts to complete a job without all the financial tools in place to do so.

The step of clarifying and communicating the scope of work is the next critical work area fraught with risk. The safest approach to scope of work is to clearly define what you have included while stipulating that you will handle anything outside the scope of work with a separate quote and signed change order. Limiting risk on materials by using allowances, the phrase "supplied by others," or a well-defined, prepriced item can make a line item loss a fairly rare occurrence.

If you look for these risk minefields, you can plan and execute strategies to minimize risk or transfer the risk to others. A poorly defined scope of work is the largest risk in labor or materials quotes. The scope of work must make crystal clear what is included and what is not. Any number of inherent "gray" areas overlap in the trades. For example, the responsibility for setting steel columns in a basement rest with the mason, the framer, or another tradesperson.

Additionally, the responsibility for installing the duct for a range hood may fall to the trade contractor for heat, ventilation, and air-conditioning

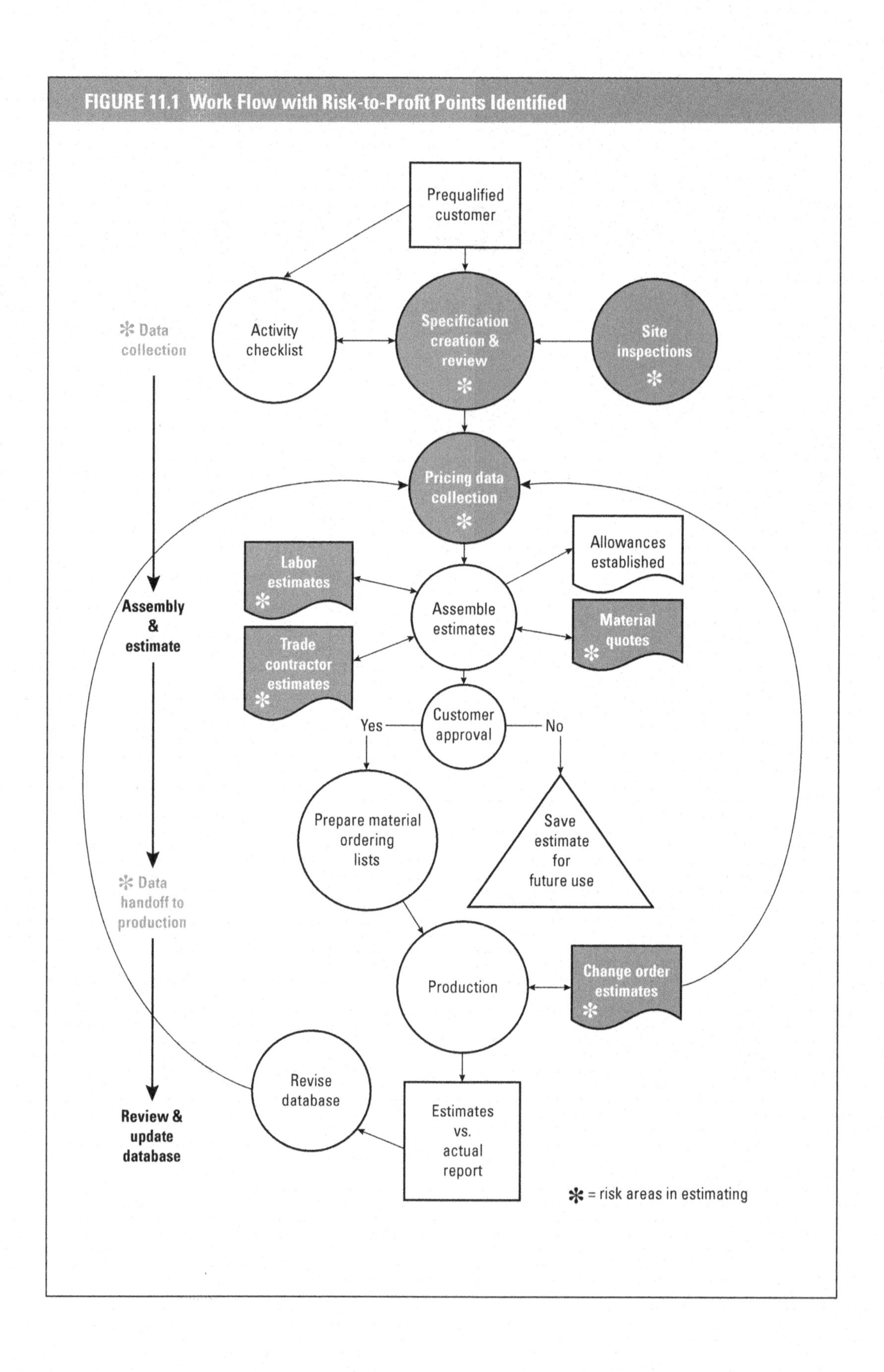
FIGURE 11.1 Work Flow with Risk-to-Profit Points Identified
Prequalified customer
Activity checklist
Specification creation & review
Site inspections
Data collection
Pricing data collection
Labor estimates
Allowances established
Assemble estimates
Material quotes
Trade contractor estimates
Assembly & estimate
Customer approval
Yes
No
Prepare material ordering lists
Save estimate for future use
Data handoff to production
Production
Change order estimates
Revise database
Estimates vs. actual report
Review & update database
= risk areas in estimating

(HVAC), the electrician, the cabinet installer, or the supervisor. A lot of finger pointing can occur on the jobsite, or the office may have to handle back charges when you don't fully define the scope of work. If the estimator omits minor pieces of the scope of work, you may have to deal with a cost overrun to get the work done. Every time you discover one of these seemingly minor items, you should revise the core estimating spreadsheet so this item does not reappear as a surprise another day.

A disciplined strategy to protect profits is a quest for clarity. Communicating well-defined expectations can minimize later frustrations of customers who expect more than the contract or your estimate includes. These defining explanations surely can take place in a site meeting or early sales conversations, but they need to be backed up in writing as an addendum to the contract for later incontrovertible clarity.

We use an addendum document that is regularly updated to include new and interesting defensive positions.

Each point comes from experience or a well-intended peer sharing valuable advice. Both parties to the contract should sign the addendum, and it becomes a part of the contract documents. Some recommended language with interpretive statements follow.

Addendum

Addendum to contract dated __________ between ______________ Building Corp. and ____________________.

1. Definitions: "__________ Building Corp. will also be referred to as the "builder"; the term "owner" includes any representative designated by the owner to act on his or her behalf in the absence of the owner; "substantial completion" means the remodeled area can be used for its intended purpose by the owner.

You can include a list of definitions so no party relies on a one-sided interpretation. For example, the term *substantial completion* means that "the owner can use the area for its intended purpose." Builders can tell plenty of stories about clients holding up final payments for items as silly as hanging a bath accessory or a back-ordered cabinet part. A $10,000 towel bar is not the intent of the contract nor is a towel bar appropriate for retainage, if retainage is necessary. If you can readily complete these incidental items, you should do so. However, the inability to perform is sometimes outside the builder's control. Therefore, the customer should make the payment when the home is substantially completed, final municipal inspections are completed, and an occupancy permit received. If work outside the scope of the contract is holding up completion, the owner should pay the balance due to be paid to the builder. At the worst, you could negotiate some amount of retainage that the customer could withhold until open issues are resolved.

2. Help us avoid misunderstandings. Any discussions or presentations that occurred prior to the signing of a contract that involves a scope of work and specifications (or any changes, additions, or deletions thereto) are preliminary in nature and usually include the offering of multiple alternatives for your consideration. From this preliminary process, an agreed-upon approach evolved for inclusion as the basis of this contract. This contract is intended to reflect only that agreed-upon approach. Unless any other alternatives and/or options are expressly spelled out within the contract document as being available for election through a certain date or milepost, they are to be deemed abandoned and not a part of the resulting contract.

Benchmarking that all prior discussion items have dissolved into the contract and plan documents is critical. A least favorite customer statement begins, "But I thought you said . . . ." After that introduction and set up, almost any of the following details could send chills down your spine. "But I thought you said . . . you would be here every day" might invoke a response such as, "No, I said the job would be supervised every day and that may or may not be by me personally. When the spackle crew is on site for a week, you do not want to pay me to watch the spackle dry. Our team is doing the best job possible with on-site work, and surely minimizing your costs for unnecessary supervision is in your best interests."

"But I thought you said . . . you could paint the garage." "Yes, the garage can be painted, but our specifications are for a tape coat only and no paint in the garage. If you want additional spackle work, priming, and painting, we can do it, but those jobs constitute additional work above and beyond the initial scope of work. May I draft a change order for your signature for that work?"

When the scope of work is well defined, alterations to it create the opportunity for a profitable change order that can address a fluid customer vision of the perfect project. Inevitably the changing phases of the work solidify the impression of the completed project versus the perceptions from the plans. You should address each instance of this added clarity (change in scope) as a new separate job in the form of a signed change order rather than turning it into a gratuity paid from a builder's profits.

"But we talked about storage space under the eaves." "Yes, we talked about many things during the formative discussions, but many of the ideas were not incorporated in the final plans when you chose among many competing ideas. If you would like to alter the scope of work now to include additional storage space, we can accommodate that with a change order."

The sample language offered in Clause 2 addresses the clearing of all prior ideas and consolidates consensus into the contracted scope of work. This sample language is a great way to put a stop to the phrase, "but I thought you said . . ." ringing in your ears.

3. Our contract pricing and construction schedule anticipate that, once the work has commenced, we will have continuous access to the work site and be able to perform our work in a continuous manner. Work stoppage and discontinuous workflow requested by the owner will create additional costs and, therefore, will result in a change order billing commensurate with the additional expenses incurred.

This clause would generally not apply for tract homes or for a custom home on a distant lot, but with so many knock-downs and infill properties being constructed, it is an easy additive to any builder's contract. Many towns have quiet ordinances that limit the hours of operation for construction, and home owners chime in with their own set of constraints. Clients occasionally want to shut down a job when they go on vacation. They may be afraid of loss of personal control or inability to perform personal inspections, but they also want the builder to know who is in control. An addendum clause addresses this situation up front by stipulating that the builder gets access to the site once the contract is in force.

4. Unless otherwise provided by the builder, all work completed by the builder will be warranted for one year from the date of its completion. All claims to the builder must be made in writing within one year from the date of substantial completion (see warranty document).

State law often impacts the duration and details of a warranty. You need to include in the contract and the warranty document some stipulation as to length, a trigger date for the start of the warranty, and a process for handling warranty work. You should make some accommodation in the original estimate to anticipate the occasional warranty request and that the line item pays for a minimum service call. Warranty service is not a line item that needs to be widely discussed, but surely you will need to go back to the home for something. Including the cost of that service call in the original estimate is imperative.

5. The builder's one-year warranty to the owner is voided if the customer fails to make any payment on the contract or fails to make payment on any change order or the builder, in exercising his or her rights hereunder, stops the work before it is completed.

The warranty has value, and withholding its application may be the only construction leverage a builder has when deductions appear on a list with a final payment that is short some dollars.

6. All warranties are void if work performed by the builder is repaired or corrected by others. Material failure is not the responsibility of the builder.

We have had clients retain money at the end of the job for their paintwork or their sweeping, or their cleaning of the balance of the house, or their perceived inconvenience. I suppose they think everyone else on the job is getting paid so they should too. We do a solid defensive job on each of these production items, but invariably this situation will occur with some customers. I then explain Clause 5, which states that their actions will void the warranty on the project because they are technically in default for nonpayment. If the shortfall occurs during the job, the work will cease. I have only had to use this clause once, but I find its existence stabilizes the payment schedule.

A home owner may attempt to help by putting a pipe wrench on a dripping chromed faucet and inadvertently destroy the finish or strip the threads. A request for a new faucet under warranty may then ensue. In the case of sweat-equity contributions to a job, the owner's work can conflict with the performance on the contract. These cases in the auto industry led to the funny posters in the auto repair shop stating a fixed labor rate, then $10 per hour more if the customer watches, and $20 per hour more if the customer helped. Even the partial loss of control of the complete jobsite can cause delays and transfer risk to the builder. Clause 6 terminates the warranty items if any loss occurs in the chain of custody on the product.

You should include written manufacturer warranties with the closure of a job. Customers may challenge a specific liability transfer clause for materials supplied on a site and challenge you with a class action suit if the defective material is distributed widely enough that many are affected, but having the clause in place establishes your intent.

7. All materials and supplies furnished by the builder and not incorporated into the work hereunder, although in the owners' custody at the work site, are and remain the property of the builder. The builder has the continuing permission of the owner to remove it. In no event will such a removal result in adjustment of the contract price.

Many times customers have expected to keep all the "excess" lumber or building materials. Reassure them that some additional on-site material *saves* them money because you avoid additional trips to the lumberyard. However, these materials remain in your inventory until placed in service on the job.

8. All materials not specified shall be selected by the builder from standard materials. Deviation from standard selections and installation is not included herein and will create an additional charge.

Many builders have been caught in the uncomfortable bind of having a customer ask for a certain installation pattern for tile even though the customer knows the contract calls for "standard" installation. You could easily write off the cost of the additional work as good will, but can you do this at every line item? Can you install sod instead of planting seed at the same cost? Can you paint twice because the color is just "a little too dark"? Would you have tilt-wash windows installed when the estimate specifies standard double hung? The answer to some of these questions could be yes under the rainbow of goodwill. The cloud of deteriorating profits grows dark on the horizon for every line item when goodwill surrenders its bounty to the customer. Discipline is the watchword for doling out goodwill in defense of job profits.

We recently had a customer request a credit for two bags of insulation left on the jobsite because the customer thought the insulator clearly must have measured incorrectly and, therefore, overcharged. Consequently, the customer thought he was overpaying. In the customer's mind, these extra bags were the smoking gun that would get them a credit refund.

I stopped at the site to look at the work (and, quite frankly, to get the excess materials off the job) and found the workmanship to be excellent. The two bags were actually neatly bundled scrap materials. The installer had conscientiously cleaned up and packed to make cleanup easier for us. I removed all the debris and filed the story for future retelling. It made no sense to inform the home owner of the blunder. The customer was convinced he overpaid, and I had already covered the issue with an accurate explanation.

9. In the case of collection on this contract the owners will be responsible for legal fees, court costs, and expenses associated with nonpayment.

The owners sign this clause so that, in the case of nonpayment, they will cover your costs for collection. The courts take a far more liberal interpretation of consumer rights, so any hint by the customer of nonperformance by the builder may make this clause irrelevant. When both sides assert claims, one party is unlikely to be awarded costs.

The better solution is accelerated cash flow to minimize risk. However, you should not interchange cash flow and contract profits, and nonpayment directly diminishes bottom line profits.

10. All sketches, drawings, plans, technical layouts, specifications, etc., provided by the builder will remain the property of the builder.

We have had little luck exercising this clause in court, but it sets an appropriate tone for nonpayment. Winning on a collection issue may only lead to an uncollectible lien, so I am hardened to the reality that my company will have some uncollected invoices over time.

Building plans enjoy strong protections through copyright laws. If the builder supplies these graphic documents, they should remain the builder's property. The plans represent a vision of the finished product that is owned by the builder, and he or she should retain them. Much like the tools required to build the job, the paper trail of documents on the project belong to the builder.

11. As part of this Agreement, unless otherwise specified, the builder will secure all permits and/or licenses for construction and will arrange for all appropriate inspections. The cost of permits and fees related to the performance of this contract will be a separate charge to the owners. The owners shall pay any Certificate of Occupancy requirements that are outside the scope of this contract by separate change order.

Certificate of Occupancy requirements are an entirely different category that municipalities use as a catch-all for every other thing in their ordinances that they may want done. These items are often arbitrary, not clear, not on the building plans, and based on the last seminar someone attended for continuing education. Local jurisdictions have asked for enhanced grading, sidewalk improvements, dry wells not on the construction plans, some retrofit items, and emergency egress lighting, and the next unknown request is always a mystery. Clearly you need to transfer this abyss to the owners.

12. The owner is responsible for being present for any Certificate of Occupancy inspections. If the owner cannot meet with the Certificate of Occupancy inspectors, then—provided that the owner gives the builder prompt notice (which must include a period of at least 4 business hours)—the builder will furnish an individual to await and accompany the Inspector. The builder will bill the owner under a change order for persons provided by the builder for such purposes at the rate of $60.00 per hour.

Because of uncertainties in municipal permit costs, we have recently moved to directly billing permit costs to the customer with an internal line item for time to prepare and get the permits.

The entire inspection process has a major impact on construction scheduling. When inspections are concurrent with other work taking place on the site, supervision is a productive cost item. However, costs are involved in opening a jobsite, confirming the inspector's approval, or fully understanding items the inspector may fail when the job is complete with no work left to do. This equivalent of house sitting is worth the cost of the opportunity to be working on something else somewhere else.

You could easily add a few hundred dollars to the estimate up front as a line item, but the reality of getting a job signed with all costs included and accounted for tends to make the efficient person the highest bidder. The adage in Chapter 1, that the single largest mistake in estimating is omission, could be countered by the possibility that the estimator who adds a cost for every contingency may not ever get the job. The balance to this scenario is to move the risk elsewhere and defend the numbers and scope of work that are communicated in your proposal. Clause 12 removes the risk from your estimate and allows for billing to the client if the job cost appears.

13. The builder maintains adequate insurance for the practice of construction, and the builder will provide certificates of insurance to the owner upon request. The owner is advised to review the current fire and liability insurance on his or her home and confirm with the agent that it is commensurate with the risks. The owner must add the builder's or the building company's name to the home owner's policy as an additional party with insured risks. The owner stands reminded that with any improvement or addition, the value of replacement goes up. The owner must be sure to increase his or her home owner's insurance policy as needed to cover the increased property value. Carrying adequate insurance is mandatory, and you should advise your client to notify his or her agent of the changing responsibilities and insured policy limits. Remember, as soon as materials are delivered to the owner's property, they are insured under the owner's home owner's policy because they are in the owner's custody, and are on, or have become a part of, the owner's property.

We had a client assert that the half-inch plumbing pipes stubbed out for roughing in a kitchen sink were a hazard to their kids playing in the unfinished kitchen. This project was one of several that have made their way to court in which we have been able to prevail.

Number 13 contains no magic; it is simply a statement of insurance coverage needed. Adding the builder as another party with insured interest may cover items that are not covered by a builder's risk policy as they are installed in a home under construction, and the builder addresses his or her own interests. A prudent home owner also needs to keep the fire policy amounts current as the job progresses.

14. In the event that a scheduled payment is more than three days past due, the builder reserves the right to cease work on the project, declare the contract breached, and (if a contract price has been offered and accepted) to be paid for that percentage of the project already completed as well as unrealized profits for the entire project

This clause uses fairly strong language in identifying a breach of contract with a specific number of days delinquent on payments. Three days are finite. Even with the benefit of the doubt about working days versus calendar days, the days are final, and the contract is in breach. Once breach is asserted, you have a number of recourse actions available, and you should notify your client of these early.

Getting your money is a serious risk in the building business, and managing the risk by curtailing the time to breach can significantly reduce the risk. Without a control date for payments, you have no way to sunset a waiting period to be paid. Confusion, uncertainty, and excuses disappear when finality is written into the earliest possible conversations and the paper trail.

As was stated earlier in a discussion of progress payments, including a contribution to your overhead and profit as you go forward, should be billable in a breach. You have committed labor and time to a job schedule; a breach on an existing contract can be devastating to work flow and profits if you are not compensated. This clause sets in place an agreement to pay within three calendar days and the right to recover profits that would accompany the job.

Early in my career I worked for a "Wednesday/Friday builder." He would tell his suppliers and trade contractors, after much delinquency, that they would be paid in full on Wednesday. Anyone can wait until Wednesday. Then on Wednesday, he would create some reason for delay to reassure them that the checks would be ready on Friday. Anyone can wait until Friday. On Friday, some emergency would arrive (all sorts of emergencies occurred over time), and he would tell them that the checks would be available on Wednesday. Anyone can wait until Wednesday. And so the cycle would continue for weeks on end. The beauty of this scam was that Wednesday and Friday never came, so the deadline was an exasperating cycle. You need to take care not to enter this cycle with a home owner who simply has no intent to honor contract responsibilities.

> 15. The contract will be satisfied upon substantial completion of all work per plans, drawings, and the agreement; final inspection by the governing bodies; or use of the space for which it was intended.

Substantial completion has a number of perspectives, and you need to address each of them early. A client may think the job is substantially complete after the last ink has run out on a series of ever-pickier complaint lists. A municipality may define substantial completion as the issuance of a Certificate of Occupancy. The municipality's interests may include that long list of noncontract issues that get attached to your permits. Clause 15 sets out a definition of substantial completion: if all the inspections are completed or if the space is being used for the purpose for which it was intended, it is substantially complete and payments are due. This language at least states that your final payment should be collected and that customer requests flow into warranty work that should not hold up payment.

> 16. All modifications/changes to the existing plans/drawings must be made in writing upon a form provided by the builder. Said form must be signed by both the owner and the builder. (This clause will not apply to projects paid on a time-and-materials basis). Only one signature from each respective party shall be necessary to execute a change order. Deviations from standard materials, procedures, or performance not included herein will result in additional charges.

When pricing is driven by scope of work and specifications, if either scope or specifications change, so must the price change. This common sense escapes most consumers who may not care what you have in the proposal so long as they get the work done their way, ideally without an up-charge or change order. The clarity and importance of this clause first surfaced with tile installations. Clients would select a tile within a square-footage budget and then expect it on a diagonal or want colors woven into a pattern on the wall or floor. This nonstandard installation would cost more, and the customer wanted it at the same contract price. The importance of benchmarking scope of work and specifications is now absolutely clear and, in fact, is more important than the price.

Price is almost irrelevant when scope is left open. Even when defined by a set of plans, delivery and technique further impact price and separate the bidders by the service they provide.

This exercise is a wonderful way to separate the true price shoppers from those who are seeking to extract value within a reasonable set of performance guidelines. A change in scope, specifications, or plans needs a signed change order that carries a price to compensate the builder for the work and risk.

You can find a nearly risk-free business relationship in cost-plus contracts. The scope and performance standards can still be in conflict between the parties as long as (a) supervision is a line item and (b) costs are tracked accurately. Then materials are tracked through job costing to create the least risky contract—a cost-plus agreement. The "plus" generally runs from 10% to 25% and covers overhead, administration, purchasing, jobsite quality control, and compliance inspections.

When faced with price-sensitive clients, I state that we can do any job they want for almost any price as long as I control the scope of work. They are now really intrigued about how far to push this price demand so I tell them that I can do their 20 × 20–foot addition for $1,000. Now with the customers in complete disbelief, I go on to describe a blue tarp over a rope stretched into the back yard. I tell them, "There's your roof and walls, interior space, room for furniture, exterior weather proofing, and a really fast construction sequence." Point made. Now I ask if they clearly understand scope of work and how it drives price.

17. In the event that the builder is entitled to a change order under any of the provisions of this contract and the owner does not come to agreement with the builder as to the terms and pricing of such a change order, or if the owner fails or refuses to make any contract or change order payments, the builder shall have the right to stop all work until the situation is fully remedied, and the owner has unconditionally reaffirmed all of his or her obligations hereunder.

Similar to Clause 16 in which a breach of the original contract is detailed, this wordage ties the breach to failure to sign change orders that may be due as a result of unexpected conditions. Our standard change order form language (found in *The Paper Trail: Systems and Forms for a Well-Run Remodeling Company* by William Asdal and Wendy A. Jordan),[5] details that "absent a fixed price, any change in scope, specifications, or installation that deviates from industry standards shall be billed on a time and materials basis." This wordage gives a framework for compensation at the minimum. Billing hourly may not be a home-run scenario for a successful builder, but at least the meter is running forward and not in reverse on profits.

We have often found and repaired bad work or unforeseen conditions without a charge to the owners, but doing so is clearly a gift. In the case of an outright gift of time or materials, you should write up a no-charge change order anyway to ensure the principal that any deviations from the plans shall be in writing and further clarify that they should cost money to correct.

18. The builder is not responsible for existing, concealed conditions that may be revealed during construction. If any existing, concealed conditions interfere with completion of this project, they will be cured on a time-and-materials basis per the contract. The owner will be advised of same as soon as practically possible.

This foundational clause follows the definition of the scope of work. Anything that was not seen during the site inspection or is later revealed that will require remediation will be completed and billed on a time-and-materials basis.

Without signed change orders and regular billing, a client can assert that the jobsite practice was to do these repairs without cost, and the precedent is set for further nonpayment. The moral here is that the magic is not in the contract, its terms, or conditions, but in the builder's team's ability to execute a professional agreement and the team's discipline to work as the contract directs.

19. Starting and completion dates are estimates, and they can be affected by weather, material shortages, changes to the proposed plans, etc. The intent of the builder is to proceed on this project through completion as expeditiously as practical.

A standard contract will "index" the start date to the issuance of permits or some other critical triggers. For example, one states that work will begin within 10 days of the issuance of municipal permits. If the permits are held up for weeks or months, the contract can still be valid and the work started within a window after issuance. With the completion date depending on the start date (that is, 14 weeks to completion), this indexing gives both parties a window during which the work gets done, but it does not put a hard date into the calendar.

20. Handling of regulated or hazardous building materials (i.e., asbestos, lead-based paint, etc.) is not included in this Agreement.

The work of handling specialized material can be an uncontrollable expense and a high risk to profits. You would do best to separate it from the scope of work with this clause and bill it on a time-and-materials basis. Bringing in specialists for mitigation can also transfer later risk should you have a claim against the job for health issues.

Inspectors have asked us for additional and excessive fire-blocking, plumbing upgrades, grading issues, testing, and electrical requirements that exceed code and are not on the plans, and we are expected to complete them to keep the job running. These costs need to be billable to others because you would have no idea up front of what these spontaneous requirements might be.

21. Any changes or alterations of the plans or specifications required by any public body, inspector, or private or governmental agency shall constitute additional work whose cost will be borne by the owner. Any inspections (engineering, compaction, soils testing, water testing, environmental, etc.) beyond municipal officials' oversight as required by a governing body shall be at additional cost.

Most often the builder and the client think that they define the scope. This statement is not entirely true, because the municipality often chimes in with a demand or three that require additional work. Clause 21 identifies and isolates these costs to the home owner.

22. Unpaid balances 30 days past due will be billed with interest charges at the rate of 1½% per month.

Clause 22 starts an interest-rate meter on bills that are past due. A usury ceiling controls the rate that you can charge, and this clause sets a legal rate that can be billed and collected.

23. All workmanship shall conform to the guidelines found in the book *Residential Construction Performance Guidelines for Professional Builders & Builders,* 3rd edition, published by BuilderBooks, of the National Association of Home Builders, 2005. If an item is not covered in that publication, standard industry practice shall govern.

This book (*Residential Construction Performance Guidelines*) is my favorite.[6] Attach this book to your contract, and you incorporate the full text into the contract as performance guidelines. More than once, we have been challenged to defend "poor drywall work" that is only visible with a 500-watt halogen light tight to the wall to create the perfect shadows, or to visit a site during the morning, only to be told the problem is only visible after 4 p.m. when the setting sun is just right. The guidelines define an acceptable level of workmanship, and in my opinion, this set of guidelines is the only one in the industry that does justice to defining workmanlike performance.

24. During the course of excavation, if any conditions are found requiring blasting or mechanical removal of rock or hardpan materials, this work will be billed as a change order, and the cost will be in excess of assumed normal excavation. Additional work could include jack hammering, drilling, and expansion cones or blasting.

Every contract doing excavation needs a blasting clause. When working with soils, you have little chance to know with certainty just what will be found below the ground. You could encounter bedrock, tight shale, and water tables requiring nearly constant pumping. These conditions are not found on the site plan or on the blueprints, so their existence can be a billable recovery of cost.

25. All work performed by the builder will be supervised by a designated representative, who may or may not be on site during the completion of the project.

This homespun language addresses the home owners who thought they had bought me for four months to sit at their site through completion. I had not previously well defined who was supervising what and when, so this clause, at least, says we may not be on site all the time. More than one customer through the years has commented that it was a nice day and nobody was at the jobsite. Any builder knows that you sometimes have days between trades when the trades are awaiting materials or you are waiting for an inspection. At such times being on the site running expenses for little progress is inefficient.

26. All sizes and dimensions of lumber or millwork will be expressed in nominal dimensions.

Rare though they may be, you may still have customers looking for full-thickness materials (1-inch hardwood is ¾-inch thick, a 2×4 is

$1\frac{9}{16} \times 3\frac{5}{8}$ inches, 1×6 clapboard is $\frac{3}{4} \times 5\frac{1}{4}$ inches, and ½-inch plywood is $\frac{7}{16}$ inch.)

27. During the course of construction, various jurisdictions may require multiple inspections. Because the proposed work is a continuum of activity, the staff anticipates being on site to open the project and accompany the inspectors. Should these inspections be critical to work flow, and work is stopped awaiting an inspector, the estimate includes no inspector waiting time in an effort to reduce cost to the client. Any waiting time for inspectors outside a normal work flow is, therefore, an additional expense and will require a change order.

Similar to Clause 12, this one caps supervisory waiting time to a minimum (which is not stated at an hourly rate) and transfers some cost of time (at $60 per hour) to the home owner when it is dedicated to awaiting an inspection or other municipal oversight.

In case of a conflict of specifications, plans, and site conditions, the specifications shall be the first order of precedence, the plans secondary, and the site conditions follow. Any deviations from this order of precedence, which drives a change in scope of work, shall be detailed in a written change order.

Inevitably, plans, specifications, contract documents, and the municipality will not always align their instructions for the work, so a precedence clause is required to position the hierarchy of relevant writings.

Notice to Owner

1. Do not sign a contract if it is blank.
2. You are entitled to a copy of this contract at the time you sign it. Keep it to protect your legal rights.
3. The owner has the right to withdraw from this agreement without penalty within three days from the signing of this contract.

____

for ________________ Builders, LLC

Date ________________________

____

for the ______________________

____

Date ________________________

Estimated Start Date: Within 10 days of issuance of municipal permits
Estimated Completion Date: _____ weeks thereafter

You need to get these addendum terms signed along with the core proposal and drawings to complete a set of contract documents. Their incorporation into the contract documents transfers appropriate risk to others and helps to defend the estimated profits from erosion.

A few years ago a state Department of Environmental Protection commissioner was addressing a builders' group. He talked about market risks and interest rate risks to builders. He talked about environmental risks and endangered species risks and how they had changed the way business was being transacted. Each time risks have entered the builders' equations for business, the costs to the consumers have gone up, often without perceptible benefits for the consumer.

Until then, all we can do is better manage the risks at hand and try to anticipate the next risks. We must continue to create defensive positions to risk that can wipe out our profits, our companies, and our rights.

The commissioner then went on to describe his vision of the next risk to builders' businesses, that of *regulatory risk*. How right this man was in his description of a chaotic, fluid regulatory minefield. In fact, he was the cause of much of the deteriorating regulatory playing field. Rules changed often without notice, enforcement went beyond the bounds of the law, takings were by regulatory creep, and stacks of appeals that had no due date for closure continued to sit for years in a bin on someone's uninspired desk. The "time of decision" law that enables municipalities to change the ordinances right up to the time of decisions on applications was onerous by itself. It can force a complete redesign or a new application. Today, regulatory risks have no bounds on their destruction of an entrepreneurial economy. Decades or generations may pass before the pendulum of reason rebuilds some business stability and restores rights to owners and risk takers.

Defensive estimating is an early step in an overall defensive strategy for business survival. Because the landscape is fluid, so must be the clauses, terms, conditions, and systems that monitor and build security for building companies. The contents of this book are a primer for creating a state of mind for defensive postures. They are a clarion call to seek out risk where it hides and defend against its destructive power before it can strike the life out of a company and profits. To that end, happy hunting!

CHAPTER 12

Defending the Profit Line in Your Remodeling Estimate

Some components in the estimating process are inherently more risk laden than others. The graphic in Figure 12.1 is the diagram in Figure 1.2, "Work Flow in Estimating," with an overlay of potential problems that could destroy profits. Note that the figure shows no risk to profits in the process of prequalifying customers. The only danger at this point is failing to eliminate unqualified leads early enough. As a result, you may find yourself doing too many estimates and courting unqualified clients far too long before detaching from them. This situation surely impacts overhead costs because your time is valuable, and you could be doing something that would add to the profit line item for your business. The remodeler's "lead to close" time is long and expensive to maintain. You need to quickly identify and cultivate promising candidates and cull the unlikely leads.

Data collection is the first activity in which estimating can start to go wrong. If you fail to observe a critical piece of information or misread the field conditions, these oversights may begin the long road to an unprofitable job. A remodeler needs to be observant of existing conditions that can affect the scope of the work. These omissions create a deficient contract, a work in progress requiring line item costs that you may not have included, and attempts to complete a job without all the financial tools in place to do so.

The step of clarifying and communicating the scope of work is the next critical work area fraught with risk. The safest approach to scope of work is to clearly define what you have included while stipulating that you will handle anything outside the scope of work with a separate quote and signed change order. Limiting risk on materials by using allowances, the phrase "supplied by others," or a well-defined, prepriced item can make a line item loss a fairly rare occurrence.

If you look for these risk minefields, you can plan and execute strategies to minimize risk or transfer the risk to others. A poorly defined scope of work is the largest risk in labor or materials quotes. The scope of work must make crystal clear what is included and what is not. Any number of inherent "gray" areas overlap in the trades. For example, the responsibility for setting steel columns in a basement rest with the mason, the framer, or another tradesperson.

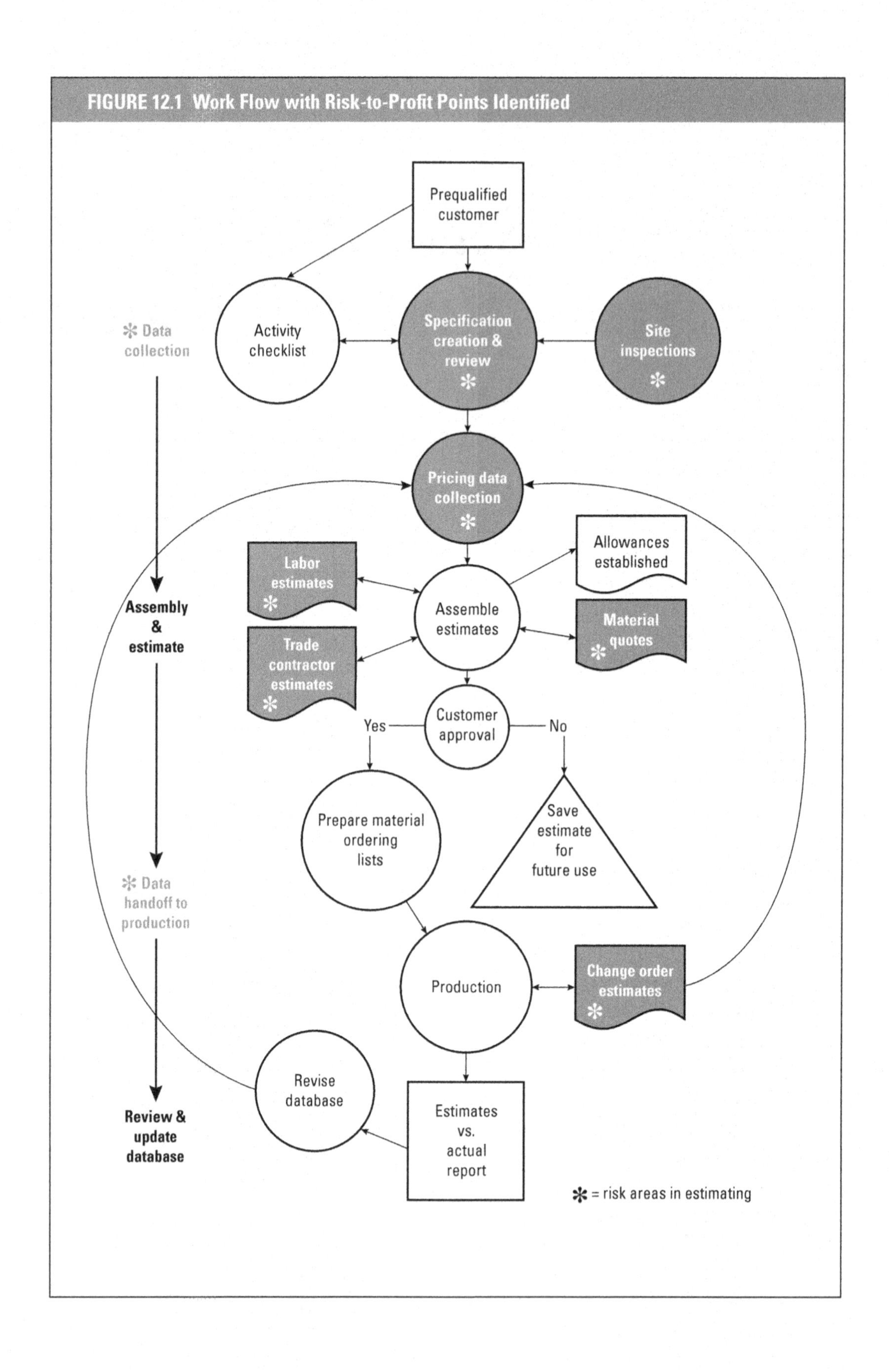
FIGURE 12.1 Work Flow with Risk-to-Profit Points Identified
Prequalified customer
Activity checklist
Specification creation & review
Site inspections
Data collection
Pricing data collection
Allowances established
Labor estimates
Assemble estimates
Material quotes
Trade contractor estimates
Assembly & estimate
Customer approval
Yes
No
Prepare material ordering lists
Save estimate for future use
Data handoff to production
Production
Change order estimates
Revise database
Estimates vs. actual report
Review & update database
= risk areas in estimating

Additionally, the responsibility for ducting a range hood may fall to the trade contractor for heat, ventilation, and air-conditioning (HVAC), the electrician, the cabinet installer, or the supervisor. A lot of finger pointing can occur on the jobsite, or the office may have to handle back charges when you don't fully define the scope of work. If the estimator omits minor pieces of the scope of work, you may have to deal with a cost overrun to get the work done. Every time you discover one of these seemingly minor items, you should revise the core estimating spreadsheet so this item does not reappear as a surprise another day.

A disciplined strategy to protect profits is a quest for clarity. Communicating well-defined expectations can minimize later frustrations of customers who expect more than the contract or your estimate includes. These defining explanations surely can take place in a site meeting or early sales conversations, but they need to be backed up in writing as an addendum to the contract for later incontrovertible clarity.

We use an addendum document that is regularly updated to include new and interesting defensive positions.

Each point comes from experience or a well-intended peer sharing valuable advice. Both parties to the contract should sign the addendum, and it becomes a part of the contract documents. Some recommended language with interpretive statements follow.

Addendum

Addendum to contract dated ___________ between _______________ Remodeling Corp. and ____________________.

1. Definitions: "___________ Remodeling Corp. will also be referred to as the "remodeler"; the term "owner" includes any representative designated by the owner to act on his or her behalf in the absence of the owner; "substantial completion" means the remodeled area can be used for its intended purpose by the owner.

You can include a list of definitions so no party relies on a one-sided interpretation. For example, the term *substantial completion* means that "the owner can use the area for its intended purpose." Remodelers can tell plenty of stories about clients holding up final payments for items as silly as hanging a bath accessory or a back-ordered cabinet part. A $10,000 towel bar is not the intent of the contract nor is a towel bar appropriate for retainage, if retainage is necessary. If you can readily complete these incidental items, you should do so. However, the inability to perform is sometimes outside the remodeler's control. Therefore, the customer should make the payment when the results of the remodeling job go into service or final municipal inspections are completed for work done by the remodeler. If work outside the scope of the contract is holding up completion, the owner should pay the remodeler's balance due. At the worst, you could negotiate some amount of retainage that the customer could withhold until open issues are resolved. A comfort level is often obtained at two times the economic value of the item installed until it is finished.

2. Help us avoid misunderstandings. Any discussions or presentations that occurred prior to the signing of a contract that involves a scope of work and specifications (or any changes, additions, or deletions thereto) are preliminary in nature and usually include the offering of multiple alternatives for your consideration. From this preliminary process, an agreed-upon approach evolved for inclusion as the basis of this contract. This contract is intended to reflect only that agreed-upon approach. Unless any other alternatives and/or options are expressly spelled out within the contract document as being available for election through a certain date or milepost, they are to be deemed abandoned and not a part of the resulting contract.

Benchmarking that all prior discussion items have dissolved into the contract and plan documents is critical. A least favorite customer statement begins, "But I thought you said . . . ." After that introduction and set up, almost any of the following details could send chills down your spine. "But I thought you said . . . you would be here every day" might invoke a response such as: "No, I said the job would be supervised every day and that may or may not be by me personally. When the spackle crew is on site for a week, you do not want to pay me to watch the spackle dry. Our team is doing the best job possible with on-site work, and surely minimizing your costs for unnecessary supervision is in your best interests."

"But I thought you said . . . you could paint the garage." "Yes, the garage can be painted, but our specifications are for a tape coat only and no paint in the garage. If you want additional spackle work, priming, and painting, we can do it, but those jobs constitute additional work above and beyond the initial scope of work. May I draft a change order for your signature for that work?"

When the scope of work is well defined, alterations to it create the opportunity for a profitable change order that can address a fluid customer vision of the perfect project. Inevitably the changing phases of the work solidify the impression of the completed project versus the perceptions from the plans. You should address each instance of this added clarity (change in scope) as a new separate job in the form of a signed change order rather than turning it into a gratuity paid from a remodeler's profits.

"But we talked about storage space under the eaves." "Yes, we talked about many things during the formative discussions, but many of the ideas were not incorporated in the final plans when you chose among many competing ideas. If you would like to alter the scope of work now to include additional storage space, we can accommodate that with a change order."

The sample language offered in Clause 2 addresses the clearing of all prior ideas and consolidates consensus into the contracted scope of work. This sample language is a great way to put a stop to the phrase, "but I thought you said . . ." ringing in your ears.

3. Our contract pricing and construction schedule anticipate that, once the work has commenced, we will have continuous access to the work site and be will able to perform our work in a continuous manner. Work stoppage and discontinuous work flow requested by the owner will create additional costs and, therefore, will result in a change order billing commensurate with the additional expenses incurred.

Many towns have quiet ordinances that limit the hours of operation for construction and home owners chime in with their own set of constraints. How many times has a client told you that you can't start work until a certain time of day? A regular day begins at 7 a.m., and a later start costs the crews hourly income and the job loses productivity. "But could you just wait until the kids get off to school?" The home owner may not willingly ask for a less-efficient production schedule, but that is the impact if the crews can't work to their potential. Clients occasionally want to shut down a job when they go on vacation. They may be afraid of theft, loss of personal control, or inability to perform personal inspections, but they also want the remodeler to know who is in control. An addendum clause addresses this situation up front by stipulating that we get access to the site once the contract is in force.

4. Unless otherwise provided by the remodeler, all work completed by the remodeler will be warranted for one year from the date of its completion. All claims to the remodeler must be made in writing within one year from the date of substantial completion (see warranty document).

State law often impacts the duration and details of a warranty. You need to include in the contract and the warranty document some stipulation as to length, a trigger date for the start of the warranty, and a process for handling warranty work. You should make some accommodation in the original estimate to anticipate the occasional warranty request and that the line item pays for a minimum service call. Warranty service is not a line item that needs to be widely discussed, but surely you will need to go back to the home for something. Including the cost of that service call in the original estimate is imperative.

5. The remodeler's one-year warranty to the owner is voided if the customer fails to make any payment on the contract or fails to make payment on any change order or the remodeler, in exercising his or her rights hereunder, stops the work before it is completed.

We have had clients retain money at the end of the job for their paintwork or their sweeping, or their cleaning of the balance of the house, or their perceived inconvenience. I suppose they think everyone else on the job is getting paid so they should too. We do a solid defensive job on each of these production items, but invariably this situation will occur with some customers. I then explain Sample Language Clause 5, which states that their actions will void the warranty on the project because they are technically in default for nonpayment. If the shortfall occurs during the job, the work will cease. I have only had to use this clause once, but I find its existence stabilizes the payment schedule.

The warranty has value, and withholding its application may be the only construction leverage a remodeler has when deductions appear on a list with a final payment that is short some dollars.

6. All warranties are void if work performed by the remodeler is repaired or corrected by others. Material failure is not the responsibility of the remodeler.

A home owner may attempt to help by putting a pipe wrench on a dripping chromed faucet and inadvertently destroy the finish or strip the threads. A request for a new faucet under warranty may then ensue. In the case of sweat-equity contributions to a job, the owner's work can conflict with the performance on the contract. These cases in the auto industry led to the funny posters in the auto repair shop stating a fixed labor rate, then $10 per hour more if the customer watches, and $20 per hour more if the customer helped. Even the partial loss of control of the complete jobsite can cause delays and transfer risk to the remodeler. Clause 6 terminates the warranty items if any loss occurs in the chain of custody on the product.

You should include written manufacturer warranties with the closure of a job. Customers may challenge a specific liability transfer clause for materials supplied on a site and challenge you with a class action suit if the defective material is distributed widely enough that many are affected, but having the clause in place establishes your intent.

7. All materials and supplies furnished by the remodeler and not incorporated into the work hereunder, although in the owners' custody at the work site, are and remain the property of the remodeler. The remodeler has the continuing permission of the owner to remove it. In no event will such a removal result in adjustment of the contract price.

Many times customers have expected to keep all the "excess" lumber or building materials. Reassure them that some additional on-site material *saves* them money because you avoid additional trips to the lumberyard. However, these materials remain in your inventory until placed in service on the job.

We recently had a customer request a credit for two bags of insulation left on the jobsite because the customer thought the insulator clearly must have measured incorrectly and, therefore, overcharged. Consequently, the customer thought he was overpaying. In the customer's mind, these extra bags were the smoking gun that would get them a credit refund.

I stopped at the site to look at the work (and, quite frankly, to get the excess materials off the job) and found the workmanship to be excellent. The two bags were actually neatly bundled scrap materials the installer had conscientiously cleaned up and packed to make cleanup easier for us. I removed all the debris and filed the story for future retelling. It made no sense to inform the home owner of the blunder. The customer was convinced he overpaid, and I had already covered the issue with an accurate explanation.

8. All materials not specified shall be selected by the remodeler from standard materials. Deviation from standard selections and installation is not included herein and will create an additional charge.

Many remodelers have been caught in the uncomfortable bind of having a customer ask for a certain installation pattern for tile even though the customer knows the contract calls for "standard" installation. You could easily write off the cost of the additional work as good will, but can you do this at every line item? Can you install sod instead of planting seed at the same cost? Can you paint twice because the color is just "a little too dark"? Would you have tilt-wash windows installed when the estimate specifies standard double hung? The answer to some of these questions could be yes under the rainbow of goodwill. The cloud of deteriorating profits grows dark on the horizon for every line item when goodwill surrenders its bounty to the customer. Discipline is the watchword for doling out goodwill in defense of job profits.

9. In the case of collection on this contract the owners will be responsible for legal fees, court costs, and expenses associated with nonpayment.

The owners sign this clause so that, in the case of nonpayment, they will cover your costs for collection. The courts take a far more liberal interpretation of consumer rights so any hint by the customer of nonperformance by the remodeler may make this clause irrelevant. When both sides assert claims, one party is unlikely to be awarded costs.

The better solution is accelerated cash flow to minimize risk. However, you should not interchange cash flow and contract profits, and nonpayment directly diminishes bottom line profits.

> 10. All sketches, drawings, plans, technical layouts, specifications, etc., provided by the remodeler will remain the property of the remodeler.

We have had little luck exercising this clause in court, but it sets an appropriate tone for nonpayment. Winning on a collection issue may only lead to an uncollectible lien, so I am hardened to the reality that my company will have some uncollected invoices over time.

Building plans enjoy strong protections through copyright laws. If the remodeler supplies these graphic documents, they should remain the remodeler's property. The plans represent a vision of the finished product that is owned by remodeler, and he or she should retain them. Much like the tools required to build the job, the paper trail of documents on the project belong to the builder.

> 11. If aluminum electrical wiring is discovered, the home owner will have to pay an additional charge to bring the existing wiring up to code requirements.

This clause is consistent with one that isolates "all unforeseen conditions" as creating an additional charge. Whenever you can identify these pitfalls, you should clarify that they are not included in the contract by listing them separately in the estimate.

For example, in your generic specification for excavations, you need to state that the contract pricing assumes adequate compaction of the soils. Any deficient materials encountered will require engineering analysis and solutions that the base contract does not include. When you are remodeling, you may discover some pitfalls, such as rot behind the walls, notching of beams, asbestos tiles under a carpet, or a host of other cost-inducing challenges.

> 12. As part of this Agreement, unless otherwise specified, the remodeler will secure all permits and/or licenses for construction and will arrange for all appropriate inspections. The cost of permits and fees related to the performance of this contract will be a separate charge to the owners. The owners shall pay any Certificate of Occupancy requirements that are outside the scope of this contract by separate change order.

Certificate of Occupancy requirements are an entirely different category that municipalities use as a catch-all for every other thing in their ordinances that they may want done. These items are often arbitrary, not clear, not on the building plans, and based on the last seminar someone attended for continuing education. Local jurisdictions have asked for enhanced grading, sidewalk improvements, dry wells not on the construction plans, some retrofit items, and emergency egress lighting, and the next unknown request is always a mystery. Clearly you need to transfer this abyss to the owners.

13. The owner is responsible for being present for any Certificate of Occupancy inspections. If the owner cannot meet with the Certificate of Occupancy inspectors, then—provided that the owner gives the remodeler prompt notice (which must include a period of at least 4 business hours)—the remodeler will furnish an individual to await and accompany the Inspector. The remodeler will bill the owner under a change order for persons provided by the remodeler for such purposes at the rate of $60.00 per hour.

Because of uncertainties in municipal permit costs, we have recently moved to directly billing permit costs to the customer with an internal line item for time to prepare and get the permits.

The entire inspection process has a major impact on construction scheduling. When inspections are concurrent with other work taking place on the site, supervision is a productive cost item. However, costs are involved in opening a jobsite, confirming the inspector's approval, or fully understanding items the inspector may fail when the job is complete with no work left to do. This equivalent of house sitting is worth the cost of the opportunity to be working on something else somewhere else. You could easily add a few hundred dollars to the estimate up front as a line item, but the reality of getting a job signed with all costs included and accounted for tends to make the efficient person the highest bidder. The adage in Chapter 1, that the single largest mistake in estimating is omission, could be countered by the possibility that the estimator who adds a cost for every contingency may not ever get the job. The balance to this scenario is to move the risk elsewhere and defend the numbers and scope of work that are communicated in your proposal. Clause 13 removes the risk from your estimate and allows for billing to the client if the job cost appears.

14. The project site will be maintained in a safe condition and as sanitary as practical. The remodeler assumes no responsibility for any injury incurred by other persons on the project site.

Many versions of this clause are in circulation. The last thing a remodeler needs is (a) another set of untrained eyes as inspectors, (b) a client's weekly list of complaints about a remodeling job in process, (c) anyone to get hurt, or (d) the owner giving directions to the tradespeople. To keep home owners out of a new addition as long as possible, especially if they have small children and pets, some remodelers build the addition without making the opening for the door to and from the rest of the house until the addition is nearly complete. And they lock the outside door when they are not working.

Carrying adequate insurance is mandatory, and you should advise your client to notify his or her agent of the changing responsibilities and insured policy limits.

We had a client assert that the half-inch plumbing pipes stubbed out for roughing in a kitchen sink were a hazard to their kids playing in the unfinished kitchen. This project was one of several that have made their way to court in which we have been able to prevail.

> 15. The remodeler maintains adequate insurance for the practice of construction and remodeling work, and the remodeler will provide certificates of insurance to the owner upon request. The owner is advised to review the current fire and liability insurance on his or her home and confirm with the agent that it is commensurate with the risks. The owner must add the remodeler's or the remodeling company's name to the home owner's policy as an additional party with insured risks. The owner stands reminded that with any improvement or addition, the value of replacement goes up. The owner must be sure to increase his or her home owner's insurance policy as needed to cover the increased property value. Remember, as soon as materials are delivered to the owner's property, they are insured under the owner's home owner's policy because they are in the owner's custody, and are on, or have become a part of, the owner's property.

Number 15 contains no magic; it is simply a statement of insurance coverage needed. Adding the remodeler as another party with insured interest may cover items that are not covered by a remodeler's risk policy as they are installed in a remodeling job, and the remodeler addresses his or her own interests. A prudent home owner also needs to keep the fire policy amounts current as the job progresses.

> 16. In the event that a scheduled payment is more than three days past due, the remodeler reserves the right to cease work on the project, declare the contract breached, and (if a contract price has been offered and accepted) to be paid for that percentage of the project already completed as well as unrealized profits for the entire project

This clause uses fairly strong language in identifying a breach of contract with a specific number of days delinquent on payments. Three days are finite. Even with the benefit of the doubt about working days versus calendar days, the days are final, and the contract is in breach. Once breach is asserted, you have a number of recourse actions available, and you should notify your client of these early.

Getting your money is a serious risk in the remodeling business, and managing the risk by curtailing the time to breach can significantly reduce the risk. Without a control date for payments, you have no way to sunset a waiting period to be paid. Confusion, uncertainty, and excuses disappear when finality is written into the earliest possible conversations and the paper trail.

As was stated earlier in a discussion of progress payments, including a contribution to your overhead and profit as you go forward, should be billable in a breach. You have committed labor and time to a job schedule; a breach on an existing contract can be devastating to work flow and profits if you are not compensated. This clause sets in place an agreement to pay within three calendar days and the right to recover profits that would accompany the job.

Early in my career, I worked for a "Wednesday/Friday builder." He would tell his suppliers and trade contractors, after much delinquency, that they would be paid in full on Wednesday. Anyone can wait until Wednesday. Then on Wednesday, he would create some reason for delay to reassure them that the checks would be ready on Friday. Anyone can wait until Friday. On Friday, some emergency would arrive (all sorts of emergencies occurred over time), and he would tell them that the checks would be available on Wednesday. Anyone can wait until Wednesday. And so the cycle would continue for weeks on end. The beauty of this scam was that Wednesday and Friday never came, so the deadline was an exasperating cycle. You need to take care not to enter this cycle with a home owner who simply has no intent to honor contract responsibilities.

17. The contract will be satisfied upon substantial completion of all work per plans, drawings, and the agreement; final inspection by the governing bodies; or use of the space for which it was intended. Any of these three events mandates the final payment.

Substantial completion has a number of perspectives, and you need to address each of them early. A client may think the job is substantially complete after the last ink has run out on a series of ever-pickier complaint lists. A municipality may define substantial completion as the issuance of a Certificate of Occupancy. The municipality's interests may include that long list of noncontract issues that get attached to your permits. Clause 17 sets out a definition of substantial completion: if all the inspections are completed or if the space is being used for the purpose for which it was intended, it is substantially complete and payments are due. This language at least states that your final payment should be collected and that customer requests flow into warranty work that should not hold up payment.

18. All modifications/changes to the existing plans/drawings must be made in writing upon a form provided by the remodeler. Said form must be signed by both the owner and the remodeler. (This clause will not apply to projects paid on a time-and-materials basis). Only one signature from each respective party shall be necessary to execute a change order. Deviations from standard materials, procedures, or performance not included herein will result in additional charges.

When pricing is driven by scope of work and specifications, if either scope or specifications change, so must the price change. This common sense escapes most consumers who may not care what you have in the proposal so long as they get the work done their way, ideally without an up-charge or change order. The clarity and importance of this clause first surfaced with tile installations. Clients would select a tile within a square-footage budget and then expect it on a diagonal, or want colors woven into a pattern on the wall or floor. This nonstandard installation would cost more, and the customer wanted it at the same contract price. The importance of benchmarking scope of work and specifications is now absolutely clear and, in fact, is more important than the price.

When faced with price-sensitive clients, I state that we can do any job they want for almost any price as long as I control the scope of work. They are now really intrigued about how far to push this price demand so I tell them that I can do their 20 × 20–foot addition for $1,000. Now with the customers in complete disbelief, I go on to describe a blue tarp over a rope stretched into the back yard. I tell them, "There's your roof and walls, interior space, room for furniture, exterior weather proofing, and a really fast construction sequence." Point made. Now I ask if they clearly understand scope of work and how it drives price.

Price is almost irrelevant when scope is left open. Even when defined by a set of plans, delivery and technique further impact price and separate the bidders by the service they provide.

This exercise is a wonderful way to separate the true price shoppers from those who are seeking to extract value within a reasonable set of performance guidelines. A change in scope, specifications, or plans needs a signed change order that carries a price to compensate the builder for the work and risk.

You can find a nearly risk-free business relationship in cost-plus contracts. The scope and performance standards can still be in conflict between the parties as long as (a) supervision is a line item and (b) costs are tracked accurately. Then materials are tracked through job costing to create the least risky contract—a cost-plus agreement. The "plus" generally runs from 10% to 25% and covers overhead, administration, purchasing, jobsite quality control, and compliance inspections.

19. In the event that the remodeler is entitled to a change order under any of the provisions of this contract and the owner does not come to agreement with the remodeler as to the terms and pricing of such a change order, or if the owner fails or refuses to make any contract or change order payments, the remodeler shall have the right to stop all work until the situation is fully remedied, and the owner has unconditionally reaffirmed all of his or her obligations hereunder.

Similar to Clause 16 in which a breach of the original contract is detailed, this wordage ties the breach to failure to sign change orders that may be due as a result of unexpected conditions. Our standard change order form language (found in *The Paper Trail: Systems and Forms for a Well-Run Remodeling Company* by William Asdal and Wendy A. Jordan),[7] details that "absent a fixed price, any change in scope, specifications, or installation that deviates from industry standards shall be billed on a time and materials basis." This wordage gives a framework for compensation at the minimum. Billing hourly may not be a home-run scenario for a successful remodeler, but at least the meter is running forward and not in reverse on profits.

> 20. The remodeler is not responsible for existing, concealed conditions that may be revealed during construction. If any existing, concealed conditions interfere with completion of this project, they will be cured on a time-and-materials basis per the contract. The owner will be advised of same as soon as practically possible.

This foundational clause follows the definition of the scope of work. Anything that was not seen during the site inspection or is later revealed that will require remediation will be completed and billed on a time-and-materials basis.

Without signed change orders and regular billing, a client can assert that the jobsite practice was to do these repairs without cost, and the precedent is set for further nonpayment. The moral here is that the magic is not in the contract, its terms, or conditions, but in the remodeler's team's ability to execute a professional agreement and the team's discipline to work as the contract directs.

> 21. The owner agrees to allow the remodeler to place a sign on the front of the premises for purposes of identification while construction is in progress.

This clause is not particularly defensive in nature, but it allows the remodeler to identify the site with a job sign. This sign facilitates deliveries and arrivals to the job. As an aside, 93% (National Association of Home Builders Survey of Remodeling Practices)[8] of all remodeling jobs are awarded through reference and referral so knowing the neighbors can be a great way to stay flush with contacts for new work.

We have often found and repaired bad work or unforeseen conditions without a charge to the owners, but doing so is clearly a gift. In the case of an outright gift of time or materials, you should write up a no-charge change order anyway to ensure the principal that any deviations from the plans shall be in writing and further clarify that they should cost money to correct.

22. Starting and completion dates are estimates, and they can be affected by weather, material shortages, changes to the proposed plans, etc. The intent of the remodeler is to proceed on this project through completion as expeditiously as practical.

A standard contract will "index" the start date to the issuance of permits or some other critical triggers. For example, one states that work will begin within 10 days of the issuance of municipal permits. If the permits are held up for weeks or months, the contract can still be valid and the work started within a window after issuance. With the completion date depending on the start date (that is, 14 weeks to completion), this indexing gives both parties a window during which the work gets done, but it does not put a hard date into the calendar.

23. Handling of regulated or hazardous building materials (i.e., asbestos, lead-based paint, etc.) is not included in this Agreement.

The work of handling specialized material can be an uncontrollable expense and a high risk to profits. You would do best to separate it from the scope of work with this clause and bill it on a time-and-materials basis. Bringing in specialists for mitigation can also transfer later risk should you have a claim against the job for health issues.

24. Any changes or alterations of the plans or specifications required by any public body, inspector, or private or governmental agency shall constitute additional work whose cost will be borne by the owner. Any inspections (engineering, compaction, soils testing, water testing, environmental, etc.) beyond municipal officials' oversight as required by a governing body shall be at additional cost.

Most often the remodeler and the client think that they define the scope. This statement is not entirely true, because the municipality often chimes in with a demand or three that require additional work. Clause 24 identifies and isolates these costs to the home owner.

25. While every effort will be made to match existing materials, textures, colors, and planes, exact duplication is not assured.

This clause is a bit of a catch-all for lowering expectations. Certainly exact duplication can never be attained if the standards are raised high enough. Lighting and shadows, structural settlement, and materials availability all can be "off" by a bit and still be able to bring closure to a completed job.

Inspectors have asked us for additional and excessive fire-blocking, plumbing upgrades, grading issues, testing, and electrical requirements that exceed code and are not on the plans, and we are expected to complete them to keep the job running. These costs need to be billable to others because you would have no idea up front of what these spontaneous requirements might be.

26. Unpaid balances 30 days past due will be billed with interest charges at the rate of 1½% per month.

This clause starts an interest-rate meter on bills that are past due. A usury ceiling controls the rate that can be charged, and this clause sets a legal rate that can be billed and collected.

27. All workmanship shall conform to the guidelines found in the book *Residential Construction Performance Guidelines for Professional Builders & Remodelers,* 3rd edition, published by BuilderBooks, National Association of Home Builders, 2005. If an item is not covered in that publication, standard industry practice shall govern.

28. During the course of excavation, if any conditions are found requiring blasting or mechanical removal of rock or hardpan materials, this work will be billed as a change order, and the cost will be in excess of assumed normal excavation. Additional work could include jack hammering, drilling, and expansion cones or blasting.

Every contract doing excavation needs a blasting clause. When working with soils, you have little chance to know with certainty just what will be found below the ground. You could encounter bedrock, tight shale, and water tables requiring nearly constant pumping. These conditions are not found on the site plan or on the blueprints, so their existence can be a billable recovery of cost.

29. All work performed by the remodeler will be supervised by a designated representative, who may or may not be on site during the completion of the project.

This book (*Residential Construction Performance Guidelines*) is my favorite.[9] Attach this book to your contract and you incorporate the full text into the contract as performance guidelines. More than once, we have been challenged to defend "poor drywall work" that is only visible with a 500-watt halogen light tight to the wall to create the perfect shadows, or to visit a site during the morning, only to be told the problem is only visible after 4 p.m. when the setting sun is just right. The guidelines define an acceptable level of workmanship, and in my opinion, this set of guidelines is the only one in the industry that does justice to defining workmanlike performance.

This homespun language addresses the home owners who thought they had bought me for four months to sit at their site through completion. I had not previously well defined who was supervising what and when, so this clause, at least, says we may not be on site all the time. More than one customer through the years has commented that it was a nice day and nobody was at the jobsite. Any remodeler knows that you sometimes have days between trades when the trades are awaiting materials or you are waiting for an inspection. At such times being on the site running expenses for little progress is inefficient.

30. All sizes and dimensions of lumber or millwork will be expressed in nominal dimensions.

Rare though they may be, you may still have customers looking for full-thickness materials (1-inch hardwood is ¾-inch thick, a 2×4 is 1 9/16 × 3 5/8 inches, 1×6 clapboard is ¾ × 5¼ inches, and ½-inch plywood is 7/16 inch.)

Sample Language

31. During the course of construction, various jurisdictions may require multiple inspections. Because the proposed work is a continuum of activity, the staff anticipates being on site to open the project and accompany the inspectors. Should these inspections be critical to work flow, and work is stopped awaiting an inspector, the estimate includes no inspector waiting time in an effort to reduce cost to the client. Any waiting time for inspectors outside a normal work flow is, therefore, an additional expense and will require a change order.

Similar to Clause 13, this one caps supervisory waiting time to a minimum (which is not stated at an hourly rate) and transfers some cost of time (at $60 per hour) to the home owner when it is dedicated to awaiting an inspection or other municipal oversight.

Notice to Owner

1. Do not sign a contract if it is blank.
2. You are entitled to a copy of this contract at the time you sign it. Keep it to protect your legal rights.
3. The owner has the right to withdraw from this agreement without penalty within three days from the signing of this contract.

____

for ________________ Remodelers, LLC

Date ________________________________

____

for the ______________________________

____

Date ________________________________

Estimated Start Date: Within 10 days of issuance of municipal permits
Estimated Completion Date: _____ weeks thereafter

Until then all we can do is better manage the risks at hand and try to anticipate the next risks. We must continue to create defensive positions to risk that can wipe out our profits our companies, and our rights.

In case of a conflict of specifications, plans, and site conditions, the specifications shall be the first order of precedence, the plans secondary, and the site conditions follow. Any deviations from this order of precedence, which drives a change in scope of work, shall be detailed in a written change order.

Inevitably, plans, specifications, contract documents, and the municipality will not always align their instructions for the work, so a precedence clause is required to position the hierarchy of relevant writings.

You need to get these addendum terms signed along with the core proposal and drawings to complete a set of contract documents. Their incorporation into the contract documents transfers appropriate risk to others and helps to defend the estimated profits from erosion.

A few years ago a state Department of Environmental Protection commissioner was addressing a builders' group. He talked about market risks and interest rate risks to builders. He talked about environmental risks and endangered species risks and how they had changed the way business was being transacted. Each time risks have entered the builders' equations for business the costs to the consumers have gone up, often without perceptible benefits for the consumer.

The commissioner then went on to describe his vision of the next risk to builders' businesses, that of *regulatory risk*. How right this man was in his description of a chaotic fluid regulatory minefield. In fact, he was the cause of much of the deteriorating regulatory playing field. Rules changed often without notice, enforcement went beyond the bounds of the law, takings were by regulatory creep, and stacks of appeals that had no due date for closure continued to sit for years in a bin on someone's uninspired desk.

The "time of decision" law that enables municipalities to change the ordinances right up to the time of decisions on applications was onerous by itself. However, it can force a complete redesign or new applications as well. Today, regulatory risks have no bounds on their destruction of an entrepreneurial economy. Decades or generations may pass before the pendulum of reason rebuilds some business stability and restores rights to owners and risk takers.

Defensive estimating is an early step in an overall defensive strategy for business survival. Because the landscape is fluid, so must be the clauses, terms, conditions, and systems that monitor and build security for remodeling companies. The contents of this book are a primer for creating a state of mind for defensive postures. They are a clarion call to seek out risk where it hides and defend against its destructive power before it can strike the life out of a company and profits. To that end, happy hunting!

Notes

Chapter 2 Establish Company Profit Number Based on Your Income Needs

1. NAHB Economics Group and NAHB Remodelors™ Council. *The Remodelers' Cost of Doing Business Study* (Washington, D.C.: Builder Books, National Association of Home Builders, 2005), 100 pp.

Chapter 5 Seven Ways to Get the Numbers

2. Christofferson, Jay. *Estimating with Microsoft Excel*, 2nd ed. (Washington, D.C.: BuilderBooks, National Association of Home Builders, 2003).
3. ___________. *EstimatorPRO™ 5.1* (Washington, D.C.: BuilderBooks, National Association of Home Builders, 2003).

Chapter 10 Financial Analysis: Estimating the Cash Flow

4. Hanbury, Alan. "Cash Is the Gas." *Professional Remodeler*, May/June 1998, p. 68.

Chapter 11 Defending the Profit Line in Your Building Estimate

5. Asdal, William, and Wendy A. Jordan. *The Paper Trail: Systems and Forms for a Well-Run Remodeling Company* with CD (Washington, D.C.: BuilderBooks, National Association of Home Builders, 2001), pp. 146, 147, and CD. Order at www.BuilderBooks.com or call 800-223-2665).
6. NAHB Business Management & Information Technology. *Residential Construction Performance Guidelines for Builders and Remodelers* (Washington, D.C.: BuilderBooks, National Association of Home Builders, 2005); and *Residential Construction Performance Guidelines: Consumer Reference*, 3rd ed. (Washington, D.C.: BuilderBooks, National Association of Home Builders, 2005).

Chapter 12 Defending the Profit Line in Your Remodeling Business

7. Asdal, William, and Wendy A. Jordan. *The Paper Trail: Systems and Forms for a Well-Run Remodeling Company* with CD (Washington, D.C.: BuilderBooks, National Association of Home Builders, 2001), pp. 146, 147, and CD.
8. NAHB Economics Group and NAHB Remodelors™ Council. *The Remodelers' Cost of Doing Business Study* (Washington, D.C.: Builder Books, National Association of Home Builders, 2005), 100 pp.
9. NAHB Business Management & Information Technology. *Residential Construction Performance Guidelines for Builders and Remodelers*, 3rd ed. (Washington, D.C.: BuilderBooks, National Association of Home Builders, 2005); and *Residential Construction Performance Guidelines: Consumer Reference*, 3rd ed. (Washington, D.C.: BuilderBooks, National Association of Home Builders, 2005).

Note: You can order items on these two pages at http://www.BuilderBooks.com or by calling 800-223-2665.

Selected Bibliography

Asdal, William, and Wendy A. Jordan. *The Paper Trail: Systems and Forms for a Well-Run Remodeling Company* with CD. Washington, D.C.: BuilderBooks, National Association of Home Builders, 2002. 306 pp.

Christofferson, Jay. *Estimating with Microsoft Excel*, 2nd ed. Washington, D.C.: BuilderBooks, National Association of Home Builders, 2003. 194 pp.

__________. *EstimatorPro 5.1*. Software. Washington, D.C.: BuilderBooks, National Association of Home Builders, 2005 (Free demo downloadable from www.builderbooks.com).

Householder, Jerry. *Estimating Home Construction Costs*, 2nd ed., with Emile Marchive III. Washington, D.C.: BuilderBooks, National Association of Home Builders, 2006. 115 pp.

NAHB Business Management and Information Technology, *NAHB Cost of Doing Business Study*. Washington, D.C.: BuilderBooks, National Association of Home Builders, 2006. 102 pp.

NAHB Economics Group and the Remodelors™ Council. *The Remodelers' Cost of Doing Business Study*. Washington, D.C.: BuilderBooks, National Association of Home Builders, 2005. 100 pp.

NAHB Remodelors™ Council and Business Management. *Residential Construction Performance Guidelines for Builders and Remodelers*, 3rd ed. (Washington, D.C.: BuilderBooks, National Association of Home Builders, 2005). 118 pp.

___________. *Residential Construction Performance Guidelines: Consumer Reference*, 3rd ed. (Washington, D.C.: BuilderBooks, National Association of Home Builders, 2005). 50 pp.

Note: You can order these products at www.BuilderBooks.com or by calling 800-223-2665.

Index

A

active income, 10–15
accuracy, 26, 76, 77–78
  in estimating production, 85
addendum to the contract
  builders, 101–13
  remodelers, 117–31
allowances, 52–54, 75
allowance, single-price, for assembly of tasks, 53, 55
aluminum electrical wiring, 122

B

bankruptcy rate, U.S., 10
barriers to estimating, 22–24
benchmark a decision, 98
breach of contract
  building, 107–8
  remodeling, 124–25
budget
  company, 9–10, 13–17
  personal, 9–15
business management, 22
business plan, 2
business, reasons to own, 5–6

C

carpenter, example of developing cost estimate, 64
carpentry, New Jersey Division of Workers' Compensation insurance rate, 87
cash flow, 4, 22, 30, 37, 39, 52
  builder's accelerated, 105
  estimating, 93–96
  remodeler's accelerated, 122
  reports, 96
  retainage clause, 96
  tips for improving, 96
cedar shakes as roof sheathing, 36
certificate of occupancy
  builder's, 106
  remodeler's, 122–23
change in lifestyle, 5, 6, 9
change orders, 79, 89
  allow work to continue without interruption, 89
  builder's, 102, 109–10, 113; breach of, 109 (*see also* breach of contract); no-charge, 110
  remodeler's, 118, 126–27, 130; breach of, 126–27 (*see also* breach of contract); no-charge, 127
chronological estimating, 30–31, 77
chronological job analysis, scheduling, 47
cleanup, 27, 97
client's needs or wants, 25–26
clients
  ideal, need to find, 69
  interview, 34–35
  Mike Turner's warning phrases about, 71
Christofferson, Jay P.
  Estimating with Microsoft Excel, 2nd, 49
  EstimatorPro™ 5.1, 50
closing ratio, industry, 75–76
communicating between jobsite and office, 87–89
communication, construction, 19–21
company profit number, 9–10, 13–17
competencies for building and remodeling success, 19–27
concrete, example of developing cost estimate, 85–86
conditions, existing/concealed
  in building, 110
  in remodeling, 127
construction scheduling
  builder's, 106
  remodeler's, 123
construction steps, 33–34
contingency cost, builder's, 107
contract documents
  building, 113
  remodeling, 131
contract prices and performance, 56, 57
contracts
  cost-plus: builder's, 109; remodeler's, 126
  fixed-price, 79
  home owners' right to withdraw: from building contract, 113; remodeling contract, 131
  project-management fee, 73–74
  standard, indexed start and finish dates: builder's, 110; remodeler's, 128
  subcontract, 65–66, 85

time-and-materials, 73, 78; builder's, 110; use for intricate trim job, 86; remodeler's, 127
trade, 65–66, 85
controlling costs, 77
conversational skills, importance of, 25–26
copyrighted plans
builder's, 106
remodeler's, 122
cost-plus contract
builder's, 109
remodelers, 126
costs estimates, examples for developing
carpentry, 64
concrete, 85–86
labor, 64, 65
roofing, 61–62
trim, intricate time-and-materials job, 86
costs, overhead, profit, and callback, 36
cost overruns, 86
crew, field, 96–97
custom work, 86

D

daily job log, 87–88
databases, productivity, 85
data collection, 35
builders, 99
remodelers, 115
day-rate pricing, 58
defending the profit line
for building estimate, 99–114
for remodeling estimate, 115–32
defending your profits, 75
defensive estimating, overview, 5
defensive strategies
to minimize or transfer risk: builder, 99, 101, 113; remodeler, 115, 117, 132
for early learning, 65
degree-of-difficulty multipliers, 49
delegate to others, 71–72, 74
design/build work, 70
trim, intricate time-and-materials job, 86
deviations from takeoff, 77
disclaimers
for soils conditions, 66
testing: for soil compaction, 66; for presence of radon/hazardous materials, 66
documentation/documents, 29–30, 106, 113–14
builder's addendum, 107–14
builder's warranty, 103
builder's contract, 105
contract indexes start and complete date: builder's, 110; remodeler's, 128
cost-plus contracts, 109–10
remodeler's addendum, 117–31
draw schedule, 30, 39, 93

E

electrical wiring, aluminum, 122
employee contact with client for selection decisions, 89
escalation clause, NAHB standard, 79, 82
estimate
convert to: draw or payment schedule, 37–38, 39; proposal, 36–37; schedule, 40, 47; statement, 38, 40
spreadsheet, template, 34–40, 41–47
staff discussion of, 36
versus actual: cost, 87; cost report, 96
estimates
charging for, 69–70
spreadsheet, 29–47
estimating
barriers to, 22–24
bottleneck, 70–71
chronological, 30–33
profit-centered, 1
shortcuts for, 50
systems, and pricing: allowances, 52–54; contract prices, 56–58; day-rate, 58; lump-sum, 49, 52: linear-footage, 54–55; quantity takeoff, 59; square-footage, 56
tips for improving and tracking production, 85–92
estimator's "crutches," examples, 66, 75
equipment costs, 58, 64
examples of developing cost estimates
carpentry, 64
concrete, 85–86
labor, 64, 65
roofing, 61–62
trim, intricate time-and-materials job, 86
excavation
building, 112
remodeling, 129
exclusions, 75
expectations, customers'
builder's, 101–2
remodeler's, 117–18

F

financial
analysis, 93–98
exposure, 97
management, 22
financing, availability of, 25–26
fixed-price
contract, 79
subcontract, 85

G

geographic regional multipliers, 49
guidelines, performance criteria or, 58
building, 111–12
remodeling, 129–30

H

Hanbury, Alan, Jr., 93
hazardous materials
building, 111
remodeling, 128

hedging, 83
home owners' right to withdraw from
building contract, 113
remodeling contract, 131

I

income
active, 9–10
passive, 9–15
needs, 9–14
inspections
building, 106, 113
remodeling, 122–23, 130
installations, selections and deviations from
builder's standard, 105
remodeler's standard, 121
insurance
adding builder to home owner's policy, 107
adding remodeler to home owner's policy, 124
claim consultant, 69–70
increase home buyer's policy re building materials delivered, 107
increase home owner's policy re remodeling materials delivered, 124
cast varies by employee activity, New Jersey Division of Workers' Compensation, 87

J

job analysis, chronological, 40, 47
job cost codes, 87, 97
job costing, 52, 97
builder's, 109
remodeler's, 126
job log
daily, 87–88
simplifies cost coding, 89
job-profitability report, 87
jobs, small versus large, 17
jobsite
communicating with office, 87–89
inspection, 66

L

labor
developing cost estimate, 64, 65
cost overrun, 89
costs, 64–65
in-house versus contract, 85
lead carpenter system, 64
leads
builders' unqualified, 99
remodelers' unqualified, 115
warning phrases, Mike Turner 71
legal fees, court costs, and expenses for collections
builder, 105
remodeler, 121
lifestyle and lifestyle costs, 5, 7, 9
limitations on work hours
for builders, 103
for remodelers, 119
linear-footage pricing and estimating system, 54–55
lumber or millwork, nominal dimensions
builder, 112
remodeler, 130
lumber futures, price trends, and price volitility, 79–81
lump-sum pricing and estimating system, 49, 52

M

mail, handling of, 26
management, 5
marketing and sales, 21
markup, 17–18
markup, scalable, 17
material cost overrun, 86
materials
costs, 75; monitoring for comparison to estimate, 78; control of costs, 77; market pricing and trend information, 82; match existing materials, textures, and planes, 36, 128; risks, 79
and tracking, 75–84
handling charges, 73
shortages, 83
selections and quantity, 76
specifications, 57–58
and supplies: builder's property, 104; remodeler's property, 120
takeoff, 75
minimize
risk, 97
estimating workload, 67–74
multipliers, geographic regional and degree-of-difficulty, 49

N

NAHB standard escalation clause, 79, 82
NAHB substitution of specified materials standard clause, 80–83
notice to owner
builder, 113
remodeler, 131

O

office, communicating with jobsite, 87–89
one-person team, 74
observation, 24–25
opportunity cost (delegation), 71
overhead, 69, 75
based on allowance, 35
cost to deliver a proposal, 69
and profit, 35, 37
profit, and callback costs, 36

P

passive income, 10–15
payment schedule, 37–38. *See also* draw schedule

performance criteria or guidelines, 58
building, 112
remodeling, 129–30
permits and fees
building, 106
remodeling, 122
personal budget and expenses, 9–15
personal skills
for mastering estimating, 24–27
for defending profit, 97
planning cycle, strategic, 6
planning exercise to find ideal client, 69
planning process, 22–23,
plan review/site visit, 66, 69, 85
plans, drawings, sketches, technical layouts
change in: builder's, 108–9, 111; change in remodeler's, 125–26, 128
copyrighted: builder's, 105; copyrighted remodeler's, 122
property of: builder, 103; property of remodeler, 122
precedence clause, builder, 113–14
prequalification
process like a funnel, 68
criteria, 68–69
prequalify customers, 67–71
prequalifying techniques, 67, 71
price escalation, 79–82
price volitility risk, 79
pricing strategies to minimize or transfer risk. *See also* estimating pricing and systems
builder, 99, 101, 113
remodeler, 115, 117, 132
process control, 26–27
production
benchmarking, 89
cost codes, 87
estimate becomes production schedule, 87
estimating and tracking, 85–92
and safety, 21
schedule, 86
supervision and variables, 85
supervisor, communicating expectations to the field crew, 89
work, 78
productivity, 76
databases, 85
site conditions and work specifications, 85
track by: employee, 87; work item, 87
professional, defining a, 4–5
profit
dangers to, 4
defending line item: builder's, 99–114; remodeler's, 115–32
profit margin
higher on small jobs, 17
planning process, 2, 22–23
projected, 66
protection of 2–7, 61–66, 78, 79: builder's, 101; remodeler's, 117
understanding the critical profit number, 1–7
profitability report, job, 87
projecting cash flow on the job, 93–96
proposal, 36–37
estimate of cost to deliver, 69
purchase order system, 77–78

Q

questions to help find variance action items, 96
quantity takeoff, 59–60
quiet ordinance
builders, 103
remodelers, 119

R

reconstruction process, 34
regulatory changes and restrictions, 79
regulatory risk
building, 113
remodeling, 131
relationship building with trade contractors, compensation for, 61
remodeling, rule of thumb and rule of sixes, 15
Residential Construction Performance Guidelines for Professional Builders and Remodelers, 111–12, 129–30
resources for
defending profit, 97
mastering estimating, 24–27
skills development, 23
trade contractors and vendors, 72, 74
retail pricing, 61–66
retainage clause, 96
builders, 101
remodelers, 118
review of work, 97
risk factor
earnings, 15
scaled, 15
risks
to materials costs, 79
to profit, 69, 79; assemblies, 55; minimizing, 59, 73, 97; builder's, 99–101; remodeler's, 115–17
transfer of, 75; builders, 99, 107; remodelers, 115, 123; production, 85
roofing, example of developing cost estimate, 61–62
roofing, rules of thumb and rules of sixes, 64

S

safety, production and, 21, 123–24
sales, marketing and, 21
savings needs, 14–15
savings rate, national, 10
schedule
payment, 37–38
draw, 30, 38, 39
production, by consensus, 86
selections, 89
scheduling, construction, 87
Schneider, Eric, 7

scope of work, 33, 36, 50–52, 56–57, 67–68, 78, 82, 85, 89, 97
builder's, 99–101, 107, 109, 111: change in, 109; municipalities affect on, 111
remodeler's, 115–18, 123, 126: change in, 127; municipalities affect on, 128
trades overlap: builder's, 99–101; remodeler's, 115–17
selections and installations
deviations from standard: builder's 105; remodeler's, 121
shakes as sheathing, 36
shortcuts, in estimating, 50, 66, 75, 76
sign for remodeling company, 127
six competencies for contracting success, 19–27
site conditions and work specifications, 85
site visit, 35–36
skills, personal, 24
soils conditions, disclaimers, 66
spreadsheet estimates, 29–47
templates for, 34–40, 47, 50–51
square footage pricing and estimating system, 56
staff discussion of estimate, 36
start and completion dates, indexed
builder's, 110
remodeler's, 128
statement, 38–39
steps to build the job, 33–34
strategic planning, 6, 22
strategies
early learning, 65
to minimize or transfer risk: builder, 99, 101, 113; remodeler, 115, 117, 132;
street
price, 49, 61, 64–65
subcontract, 65–66, 85
substantial completion
builders, 101, 108
remodelers, 117, 125
substitution of specified materials, NAHB standard, 80–83
supervision
cost of, 64, 73
lead carpenter system, 64
New Jersey Division of Workers' Compensation insurance rate, 87
production, 85
supervisor,
field, putting corrections into system, 96
production, 86–87, 89
communicating expectations to the field crew, 89
responsibility: builder, 100–101, 112; remodeler, 117, 129–30
supervisory waiting time
builder, 112
remodeler, 130
supply chain, 81
supply chain services, 65, 84
systems thinking, 4

T

takeoff, 59–60, 75–76,
identify deviations from, 77
team, 79
teamwork, 62, 74
templates, 30, 34–40, 47, 70
testing
for soils conditions and compaction, 66
for radon and hazardous materials, 66
"time of decision" law
for building, 114
for remodeling, 131
time management, 70, 75
time-and-materials contract and work, 73, 78
builders, 110
intricate trim job, 86
remodelers, 127
tracking
productivity, 87–92
production, estimating and, 85–92
trade contract, 65–66, 85
fixed-price, 79
installed price, 83
trade contractor, 62
acting as your own, 63
costs versus in-house labor costs, 64, 65
plan review/site visit, 65, 72, 85
schedule, 85
unit costs, 65
trade contractor work, skill level, 85
trade contractors include materials in: fixed-price contract, 79, installed price, 83
tracking expenses, 97
Turner's, Mike, warning phrases from prospects, 71

U

unit costs, trade contractor, 65
unit price, 53
unpaid balances. *See also* breach of contract
builders, 111
remodelers, 129

V

value for clients, 62, 70
variance action items, 96
vendor discounts, 96
visualization, 26

W

warranty
builder's, 103–4; manufacturer's, 104; voided by home buyer's failure to pay, 103; voided by unauthorized repairs, 104; warranty service, 103
remodeler's, 119–20; manufacturer's, 120; voided by home owner's failure to pay, 120; voided by unauthorized repairs, 120; warranty service, 119
wiring, aluminum electrical, 122
work flow, 2–4, 86,
for builders, 99
for remodelers, 115

work hours
  builders, 103
  remodelers, 119
work orders, 87
work specifications and site conditions, 85
Workers' Compensation, New Jersey Division of, 87
  reduction in costs for workers' compensation insurance, 87
  rates vary by worker activity, 87
workload, ways to lower, 70–74
  capitalize on the supply chain, 72
  delegate to others, 71–72, 74
  do fewer estimates, 69, 74
  repetition, 72
  project-management-fee contract, 73–74
  save templates, 30, 34–40, 47, 72
  time management, 70, 72–73, 75
  track turn-around time, 73
  time-and-materials contracts, 73
workmanship
  building, 111
  remodeling, 129

About the National Association of Home Builders

The National Association of Home Builders is a Washington-based trade association representing more than 225,000 members involved in home building, remodeling, multifamily construction, property management, trade contracting, design, housing finance, building product manufacturing, and other aspects of residential and light commercial construction. Known as "the voice of the housing industry," NAHB is affiliated with more than 800 state and local home builders associations around the country. NAHB's builder members construct about 80 percent of all new residential units, supporting one of the largest engines of economic growth in the country: housing.

Join the National Association of Home Builders by joining your local home builders association. Visit www.nahb.org/join or call 800-368-5242, x0, for information on state and local associations near you. Great member benefits include:

- Access to the **National Housing Resource Center** and its collection of electronic databases, books, journals, videos, and CDs. Call 800-368-5254, x8296 or e-mail nhrc@nahb.org
- **Nation's Building News**, the weekly e-newsletter containing industry news. Visit www.nahb.org/nbn
- **Extended access to www.nahb.org** when members log in. Visit www.nahb.org/login
- **Business Management Tools** for members only that are designed to help you improve strategic planning, time management, information technology, customer service, and other ways to increase profits through effective business management. Visit www.nahb.org/biztools
- **Council membership**:
  - **Building Systems Council**: www.nahb.org/buildingsystems
  - **Commercial Builders Council**: www.nahb.org/commercial
  - **Building Systems Council's Concrete Home Building Council**: www.nahb.org/concrete
  - **Multifamily Council**: www.nahb.org/multifamily
  - **National Sales & Marketing Council**: www.nahb.org/nsmc
  - **Remodelors™ Council**: www.nahb.org/remodelors
  - **Women's Council**: www.nahb.org/womens
  - **50+ Housing Council**: www.nahb.org/50plus

BuilderBooks, the book publishing arm of NAHB, publishes inspirational and educational products for the housing industry and offers a variety of books, software, brochures, and more in English and Spanish. Visit www.BuilderBooks.com or call 800-223-2665. NAHB members save at least 10% on every book.

BuilderBooks Digital Delivery offers over 30 publications, forms, contracts, and checklists that are instantly delivered in electronic format to your desktop. Visit www.BuilderBooks.com and click on Digital Delivery.

The **Member Advantage Program** offers NAHB members discounts on products and services such as computers, automobiles, payroll services, and much more. Keep more of your hard-earned revenue by cashing in on the savings today. Visit www.nahb.org/ma for a comprehensive overview of all available programs.